D1918476

The Freshwater Fishes of Europe

Vol. 9

The Freshwater Fishes of Europe

Vol. 1/I:
Petromyzontiformes
Edited by Juraj Holčík

Vol. 1/II:
General Introduction to Fishes. Acipenseridae
Edited by Juraj Holčík

Vol. 2:
Clupeidae, Anguillidae
Edited by Henri Hoestlandt

Vol. 3:
Salmonidae, Coregonidae, Thymallidae, Osmeridae, Umbridae, Esocidae
Edited by Yuriĭ Stepanovich Reshetnikov

Vol. 4 and 5:
Cyprinidae
Edited by Petru Bănărescu

Vol. 6:
Cobitidae, Siluridae, Ictaluridae
Edited by K. D. Vasilieva and V. P. Vasiliev

Vol. 7:
Gadidae, Gasterosteidae, Syngnathidae, Cyprinodontidae, Poeciliidae, Atherinidae, Cottidae, Centrarchidae, Percidae
Edited by Anton Lelek

Vol. 8:
Gobiidae, Blenniidae, Pleuronectidae
Edited by V. I. Pinchuk

Vol. 9:
Threatened Fishes of Europe
By Anton Lelek
Edited by the European Committee for the Conservation of Nature and Natural Resource–Council of Europe

If necessary, depending on the quantity of material, volumes may be divided in two parts.

The Freshwater Fishes of Europe

Vol. 9
Threatened Fishes of Europe

By Anton Lelek

Edited by the European Committee
for the Conservation of Nature and
Natural Resources – Council of Europe

with 85 figures

AULA-Verlag Wiesbaden

Research Institute Senckenberg
and Science Museum
Senckenberganlage 25
D-6000 Frankfurt/M.
and
Department of Fisheries
Institute for Wildlife Biology
and Game Management
University of Göttingen
D-3400 Göttingen

CIP-Kurztitelaufnahme der Deutschen Bibliothek
The **freshwater fishes of Europe**. – Wiesbaden:
Aula-Verlag
Vol. 9. Threatened fishes of Europe / ed. by the
Europ. Committee for the conservation of Nature
and Natural Resources, Council of Europe. By
Anton Lelek. – 1987.
ISBN 3-89104-048-2
NE: Lelek, Anton [Mitverf.]; Council of Europe /
European Committee for the Conservation of
Nature and Natural Resources

Printed in Germany / Imprimé en Allemagne
Cover design: Klaus Neumann
Typeset by Maschinensetzerei Hans Janß, Pfungstadt
Printed and bound by Druck- und Buchbinderei-Werkstätten
May + Co. Nachf. Darmstadt
ISBN 3-89104-048-2

Dedicated to my late father Antonín Lelek

Contents

1. Introduction

The primary aim of this volume is to provide a general review of freshwater fish fauna with special attention to the amount of danger to the survival of each species. Wherever possible, suggestions have been made for immediate protection of individual species. However, the actual, modern conservation has to be based on the protection of species communities and their special environments.

The actual contribution of this publication to the conservation and protection of fish species is limited by the amount of information available. For some species, the improvement of water quality and the physical characteristics of the river would be sufficient. However, little is known about the life-histories, distribution, and environmental requirements of some species. This lack of vital ecological information frequently makes it difficult to assess how endangered they are, and makes the choice of appropriate protective measures uncertain. One of prerequisites for conservation should be good zoological knowledge of the species. The efforts of ichthyologists, however, were focused mainly on descriptive somatic information about fish species, which contributed to better knowledge for taxonomy and systematics, and for improvements of nomenclature. It is hoped that this knowledge will improve the cooperation among practical-minded fishery biologists and naturalists, who are willing to contribute to actual conservation. Taking into consideration the methods used for artificial but successful management of obviously "useful" species, it is clear that much progress can be achieved by direct use or modification of standard methods for propagation of many rare, vanishing species.

It is hoped this publication will stimulate the effort to overcome the existing gaps in the knowledge of the fish communities in rivers, and particularly of their dynamics in artificially modified environments. Past experience has proved that exaggerated attention to a few "useful" species influences the remaining fish community negatively. A separation of the "utilitarian" and "conservation" usages of the aquatic environment and its fish fauna is thus called for. The present trend toward public environmental awareness, if guided by limnologists, ichthyologists, fishery biologists, and practical fishery experts, may contribute decisively to better comprehensive and effective improvements in aquatic environments, in general, and for the indigenous fauna, in particular. The fishery itself would then be based on more natural conditions and would help, wherever needed, to regulate the processes in aquatic ecosystems.

Answering questions on the threat to species and finding ways to protect them, requires much work in classical ichthyology and fish ecology. These seem to be the prerequisites for comprehensive guidelines for protection of species and their permanent conservation. The deterioration of aquatic environments is now very serious, and the outlook is gloomy. However, a compromise between possibility and necessity in environmental protection may lead to some local successes. If we whish to ensure that the ichthyofauna remains here for us and for the generations to come, radical conservation methods applied to integrated parts of catchment areas will be necessary for permanent success.

Everybody is aware of the close link between water and fish. However, human beings need water as well. When the water supply is unsuitable for drinking, it must be purified. For other purposes other treatments are necessary and the polluted remainder is left for the fish. The use of water for transportation and for the production of energy changed entire hydrological systems. The response of individual species and groups of species to alterations of the environment is dramatic. Many species have almost disappeared or are now limited to areas so small that their future existence is extremely uncertain.

Endangered fish have been given less attention than with other vulnerable terrestrial animals or plants, because they are seldom seen in the water. The vanishing of "small" species does not arouse as much concern, as the elimination of birds, butterflies, plants and mammals. No sanctuaries for fish have been established. Interest in fish has been focused on a small group of species deemed desirable, which have been artificially managed. Very little attention was usually paid to environmental demands for these species. Difficulties were overcome by sophisticated techniques of artificial breeding and stocking. This resulted in increased production per unit area of the water body and better results for the sport fishermen.

That more writing about this problem will contribute to actual improvement is doubtful. We cannot do more than repeat the words of one of the pioneers of limnological science, written over 30 years ago. Towards the end of his life THIENEMANN (1955, 1956) stated "A river system is a unity throughout its entire length and must be handled accordingly, regardless of administrative or political boundaries". Not less important was his remark: "Ecological training must penetrate the professional training of engineers" and finally his postulate remains valid: "Theoretical water research is the basis of practice and must obtain far better support . . .".

In spite of all the setbacks which we have witnessed there is some room for optimism left. The environment and the fish fauna itself have not universally deteriorated: However, if something is not done soon, all attempts to restore them may be too late.

Frankfurt/M. December 1986

2. Outline of the European freshwater fish fauna

Looking at the history of the fish fauna of Europe, one sees the decisive role played by geological changes and climatic fluctuations. The main reason that the fauna is poorer in the north-western part of Europe is that glaciation virtually eliminated the established populations. In this part of Europe, the retreat of the ice was followed by colonization from the rest of the continent. However, the glacial ice never spread southward as far as to the Alps, the Carpathians and the Pyrenees. The preglacial elements are therefore found in south-eastern Europe and in the area of the Black and the Caspian Seas.

2.1. Notes on the origin and development of the ichthyofauna in the main catchment areas

Fishes have always been bound to watercourses. It therefore is useful to provide a simple review of the present catchment areas. This can be linked with the present fish distribution and will answer some of the intriguing questions, why the fauna in some parts of Europe is richer than in others.

We must start from the late tertiary period, when Europe was connected with North America. This connection probably had little consequence for fish distribution. Only Percidae may have dispersed across a North Atlantic land route near the end of the Cretaceous or at the beginning of the Eocene period. The faunas in the east and west parts of the Eurasian continent were very probably separated by the ancient "Ob-Sea". After it dried up, some time during the late Oligocene period, came the first opportunity for the exchange of fauna in both directions. In the southern part of Europe, the "Mediterranean Sea" probably stretched from the Rhône River as far as the present Aral Sea. Its northern branch formed the "Sarmatian Sea", which was only slightly salty, thus making contact possible between Europe and Siberia. Some time near the end of the Tertiary or beginning of the Quaternary period, the "Sarmatian Sea" was divided by the rise of the Carpathian Mountains, which resulted in the formation of the Panonian Sea. Slightly salty primeval seas, including the Panonian and the Sarmatian, could probably be crossed by freshwater fish in eastward and westward directions.

The geological changes and climatic fluctuations during the Pleistocene were crucial for the area under consideration. Four major glaciations, with recessions and warmer periods between them, played their part. The second glaciation, when the ice came farthest south, was particularly important. The areas southeast of the Carpathians, as well as the Alps, provided refuges for the survival of some elements of the pre-glacial fauna. Obviously, this is the reason why the Danube catchment area is much richer than that of the Rhine or the Elbe. For detailed account about Central Europe, see THIENEMANN (1941), BĂNĂRESCU (1965) and BALON (1968). Much information on the origin of the fauna of southern Europe, particularly the Balkan Peninsula, is reviewed by STANKOVIČ (1960). The zoogeography of the Iberian fish fauna, and elements of the European fauna in North Africa, is treated by ALMAÇA (1976).

2.2. Classification systems proposed for fish distribution zones in Central Europe

It is commonly known that species distribution varies according to the physical characteristics of the stream.

Certain fish species have been limited to particular river stretches. Their names later chosen designated typical reaches of streams. They are discussed in detail in the 2nd Vol., this series. At the end of the 19th century, FRIČ (1872, 1888), BORNE (1877) and NOWICKI (1882) had already divided the streams into four basic regions: the trout, grayling, barbel and bream-ruffe zone. Later on, this classical division was improved and made more accurate by incorporating measurable physical parameters for the stream region: the slope, the width and the water temperature (HUET 1949). This classification system, suitable for the streams of Northwest Europe, was later extended to the waters of the Carpathian area of Central Europe (HUET *et al.* 1969). Some deviations from this scheme are observed in the streams of the British Isles (VARLEY 1967). There the streams are generally shorter, from the source to the estuary. This is why zonation is not as clearly developed as in continental Europe. The last and the most comprehensive classification system, which includes all aquatic organisms, was suggested by ILLIES (1961a, b). It also applies to running waters outside Europe with different, but convergent fauna. The definitive parameters in this system are the physical structure of the river bed and the range of water temperature during the year. The name of the zones that are used in this volume are listed below:

Rhithral: mountain creek and small rivulet zone

Epirhithral and metarhithral: trout zone

Hyporhithral: grayling zone

Potamal: lowland river zone

Epipotamal: barbel zone

Metapotamal and hypopotamal: bream and ruffe zone.

Both the traditional and the newest classification systems for stream sections are used in the brief descriptions of fish habitats in section 3.2, and in the section presenting the conservation proposals to avoid misunderstandings. The system does not seem to be of much importance for each particular taxon, but it may be crucial for deciding on the location of protective areas for fish taxocoens.

2.3. Provisional list of European cyclostomes and freshwater fishes

An attempt has been made to draft a list of species and some subspecies recorded from Europe. Freshwater species are considered to be those which reproduce in inland waters, the only exception being the eel. With few exceptions, brackish water species, particularly the gobies, are not trated in detail here. If the taxonomic status of a group is not fully clear, only the generally known and accepted taxa are listed. The sequence of families follows the scheme of NELSON (1976) which seems to reflect the relationships within the systematic groups according to the most up-to-date level of knowledge. The genera and species are arranged either alphabetically or according to general conventions. Internationally accepted names, as listed in the multilingual catalogue prepared by BĂNĂRESCU *et al.* (1971) and to the new summary of scientific names presented in Limnofauna Europaea, edited by ILLIES (1978) have been included. Subspecies are only mentioned with their respective species (section 3.2). In the case of species that are represented only by certain subspecies in Europe, the full subspecific name is often mentioned, and the text is focused on the European subspecies. In certain cases, the recent publications of MAITLAND (1977) and WHEELER (1978) are cited. As the manuscript of the first edition of the book, written for a report of the European Committee for the conservation of nature and natural resources of the Council of Europe, was submitted in 1979, it was beyond the possibility of the author to include all ongoing literature within this publication. The author is oblidged to the publishers for accepting changes and ammendments which, however, are far behind from being complete. Further informations on the state of threatened species will be provided in the corresponding sections within the forthcoming volumes of "The Freshwater Fishes of Europe".

The purpose of preparing this list is to demonstrate the richness of the fauna and to provide a quick guide for distinguishing indigenous species, marked with an "●", and introduced species, marked with an "○". Species marked with a "+" are treated in a particular section in the main part of this volume. The letter "**M**" indicates that a distribution map is provided; the number indicates its page number in the text. Other abbreviations used here are taken from Chapter 3, as follows: **E** = endangered, **V** = vulnerable, **R** = rare, **I** = intermediate.

Petromyzontidae		classification		page
● *Petromyzon marinus* Linnaeus, 1758	+	V		25
● *Caspiomyzon wagneri* (Kessler, 1870)	+	V		28
● *Eudontomyzon danfordi* Regan, 1911	+	V	M	29
● *Eudontomyzon hellenicus* Vladykov, Renaud, Kott and Economides, 1982	+	V		33
● *Eudontomyzon mariae* (Berg, 1931)	+	V		32
● *Lampetra fluviatilis* (Linnaeus, 1758)	+	E	M	36
● *Lampetra planeri* (Bloch, 1784)	+	V	M	39
● *Lethenteron (Lampetra) japonicum* (Martens, 1868)	+	V		34
● *Lethenteron (Lampetra) zanandreai* (Vladykov, 1955)	+	V		35

Acipenseridae				
● *Acipenser sturio* Linnaeus, 1758	+	E	M	42
● *Acipenser güldenstaedti* Brandt, 1833	+	V		46
● *Acipenser naccarii* Bonaparte, 1834–1841	+	V		47
● *Acipenser nudiventris* Lovetsky, 1828	+	V–E		48
● *Acipenser persicus* Borodin, 1897	+	E		50
● *Acipenser ruthenus* Linnaeus, 1758	+	E	M	51
● *Acipenser stellatus* Pallas, 1811	+	V		54
● *Huso huso* (Linnaeus, 1758)	+	V–E	M	55

Clupeidae				
● *Alosa alosa alosa* (Linnaeus, 1758)	+	E	M	58
● *Alosa caspia nordmanni* (Antipa, 1905)				
● *Alosa* (Caspialosa) pontica pontica (Eichwald, 1838)				
● *Alosa fallax fallax* Lacépède, 1800	+	E	M	61
● *Clupeonella cultriventris cultriventris* (Nordmann, 1840)				
Anguillidae				
● *Anguilla anguilla* (Linnaeus, 1758)	+	I–V	M	63
Esocidae				
● *Esox lucius* Linnaeus, 1758	+	I–V	M	67
Umbridae				
● *Umbra krameri* Walbaum, 1792	+	V–E	M	70
○ *Umbra pygmaea* (De Kay, 1842)			M	70
Coregonidae				
● *Coregonus albula* Linnaeus, 1758	+	E	M	75
● *Coregonus autumnalis* (Pallas, 1811) (Coregonus pollan Thompson, 1835)	+	V		78
● *Coregonus baunti* Muchomedijarov, 1948	+	V	M	79
● *Coregonus lavaretus* Linnaeus, 1758	+	E	M	81
● *Coregonus nasus* (Pallas, 1776)	+	E	M	84
● *Coregonus oxyrhynchus* Linnaeus, 1758 (*Coregonus macropthalmus* Nüsslin, 1882)	+	E	M	87
● *Coregonus peled* Gmelin, 1789 (*Coregonus muksun* Pallas, 1776)	+	E		90
● *Coregonus pidschian* Gmelin, 1789 (*Coregonus acronius* Rapp, 1854)	+	E		91
Thymallidae				
● *Thymallus thymallus* (Linnaeus, 1758)	+	V	M	93
Salmonidae				
● *Salmo salar* Linnaeus, 1758	+	E	M	97
● *Salmo trutta trutta* Linnaeus, 1758	+	E	M	101
● *Salmo t. trutta m. fario,* Linnaeus, 1758	+	V	M	105
● *Salmo trutta labrax* Pallas, 1811	+	E	M	109
● *Salmo trutta labrax* m. *fario*	+	E	M	111
● *Salmo trutta caspius* Kessler, 1877			M	106

●	*Salmo trutta lacustris* (Linnaeus, 1758)	+	E		112
●	*Salmo trutta macrostigma* (A. Dumeril, 1858)	+	V	M	114
○	*Salmo gairdneri* Richardson, 1836				
●	*Salmothymus obtusirostris* (Heckel, 1851)	+	V	M	115
●	*Salmothymus ohridanus* (Steindachner, 1892)	+	V	M	117
●	*Hucho hucho* (Linnaeus, 1758)	+	E	M	118
●	*Salvelinus alpinus* (Linnaeus, 1758)	+	E	M	122
○	*Salvelinus fontinalis* (Mitchell, 1815)				
○	*Salvelinus namaycush* (Walbaum, 1792)				
○	*Oncorhynchus gorbuscha* (Walbaum, 1792)				
○	*Oncorhynchus Kisutsch* (Walbaum, 1792)				
○	*Oncorhynchus tschawytscha* (Walbaum, 1792)				
●	*Stenodus leucichthys leucichthys* (Güldenstaedt, 1772)	+	E	M	126
●	*Stenodus leucichthys nelma* (Pallas, 1773)	+	I	M	128

Osmeridae

●	*Osmerus eperlanus* (Linnaeus, 1758)	+	V		130

Cyprinidae

●	*Cyprinus carpio* Linnaeus, 1758	+	V–E	M	132
●	*Abramis brama* (Linnaeus, 1758)				
●	*Abramis ballerus* (Linnaeus, 1758)	+	V	M	136
●	*Abramis sapa* (Pallas, 1811)	+	R–V	M	139
●	*Alburnus albidus* (Costa, 1838)	+	I	M	142
●	*Alburnus alburnus* (Linnaeus, 1758)	+	I	M	144
●	*Alburnoides bipunctatus* (Bloch, 1782)	+	V–E	M	147
●	*Alburnus charusini* Herzenstein, 1889				
●	*Aspius aspius* (Linnaeus, 1758)	+	V–E	M	150
●	*Aulopyge hügeli* Heckel, 1841	+	R	M	153
●	*Barbus barbus* (Linnaeus, 1758)	+	V	M	156
●	*Barbus meridionalis* Risso, 1826	+	V	M	160
●	*Barbus brachycephalus* Berg, 1914				
●	*Barbus capito* (Güldenstaedt, 1773)				
●	*Barbus comicza* Steindachner, 1866			M	164
●	*Barbus peloponnensius* Valenciennes, 1842			M	164
●	*Barbus plebejus* Valenciennes, 1842	+	V	M	163
●	*Blicca bjoerkna* (Linnaeus, 1758)	+	I	M	165
●	*Carassius carassius* (Linnaeus, 1758)	+	R–V	M	168
○	*Carassius auratus* (Linnaeus, 1758)	+	I	M	171
●	*Chalcalburnus chalcoides* (Güldenstaedt, 1772)	+	V–E	M	175
●	*Chondrostoma colchicum* Kessler, 1899				
●	*Chondrostoma genei* Bonaparte, 1832–1841	+	V	M	183
●	*Chondrostoma kneri* Heckel, 1843				

●	*Chondrostoma lemmingi* (Steindacher, 1866)				
●	*Chondrostoma lusitanicum* Collares-Pereira, 1980				
●	*Chondrostoma nasus* (Linnaeus, 1758)	+	I–V	M	178
●	*Chondrostoma oxyrhynchum* Kessler, 1877				
●	*Chondrostoma phoxinus* Heckel, 1843				
●	*Chondrostoma polylepis* Steindachner, 1866				
●	*Chondrostoma soetta* Bonaparte, 1832–1841	+	V	M	185
●	*Chondrostoma toxostoma* Vallot, 1837	+	V	M	181
○	*Ctenopharyngodon idella* (Valenciennes 1844)				
●	*Gobio albipinnatus* Lukasch, 1933	+	R	M	187
●	*Gobio ciscaucasicus* Berg, 1932				
●	*Gobio gobio* (Linnaeus, 1758)	+	I–R	M	190
●	*Gobio kessleri* Dybowski, 1862	+	R–V	M	194
●	*Gobio uranoscopus* (Agassiz, 1828)	+	R–V	M	197
○	*Hypophthalmichthys molitrix* (Valenciennes, 1844)				
○	*Hypophthalmichthys (Aristichthys) nobilis* (Richardson, 1845)				
●	*Iberocypris palaciosi* Doadrio, 1980				
●	*Leucaspius delineatus* (Heckel, 1843)	+	R–V	M	200
●	*Leucaspius marathonicus* (Vinciguerra, 1920)				
●	*Leucaspius stymphalicus* (Valencienes, 1844)				
●	*Leucaspius irideus* Ladiges, 1960				
●	*Leuciscus aphipsi* Aleksandrov, 1927				
●	*Leuciscus berak* (Heckel, 1843)				
●	*Leuciscus borysthenicus* (Kessler, 1859)				
●	*Leuciscus cephaloides* Battalgil, 1942				
●	*Leuciscus cephalus* (Linnaeus, 1758)				
●	*Leuciscus danilewski* (Kessler, 1877)				
●	*Leuciscus idus* (Linnaeus, 1758)	+	V–E	M	203
●	*Leuciscus illyricus* (Heckel & Kner, 1858)				
●	*Leuciscus lepidus* (Heckel, 1843)				
●	*Leuciscus leuciscus* (Linnaeus, 1758)	+	I–V	M	207
●	*Leuciscus lucumonis* Bianco, 1983				
●	*Leuciscus (Telestes) microlepis* (Heckel, 1843)				
●	*Leuciscus (Telestes) polylepis* (Steindachner, 1866)				
●	*Leuciscus (Telestes) souffia* Risso, 1826	+	V–E	M	210
●	*Leuciscus svallize* (Heckel & Kner, 1858)				
●	*Leuciscus (Telestes) turskyi* (Heckel, 1843)				
●	*Leuciscus (Telestes) ukliva* (Heckel, 1843)				
○	*Mylopharyngodon piceus* (Richardson, 1846)				
●	*Pachychilon pictum* (Heckel und Kner, 1858)				
●	*Pelecus cultratus* (Linnaeus, 1758)	+	I–R	M	213

● *Phoxinellus adspersus* Heckel, 1843				
● *Phoxinellus stymphalicus* (Cuvier & Valenciennes, 1844)				
● *Phoxinus phoxinus* (Linnaeus, 1758)	+	V	M	217
● *Phoxinus percnurus* (Pallas, 1811)	+	V–E	M	220
○ *Pimephales promelas* Rafinesque, 1820				
● *Pseudophoxinus (Phoxinellus) hispanicus* (Steindachner, 1866)				
○ *Pseudorasbora parva* (Temminck & Schlegel, 1842)				
● *Rhodeus sericeus* (Pallas, 1776)	+	R–V	M	223
● *Rutilus rutilus* (Linnaeus, 1758)				
● *Rutilus arcasii* (Steindachner, 1886)				
● *Rutilus lemmingii* (Steindachner, 1866)				
● *Rutilus macedonicus* (Steindachner, 1892)				
● *Rutilus macrolepidotus* (Steindachner, 1866)				
● *Rutilus pigus* (Lacépède, 1804)	+	R	M	226
● *Rutilus pigus virgo* (Heckel, 1852)	+	R–V	M	227
● *Rutilus rubilio* (Bonaparte, 1837)	+	I	M	230
● *Pararutilus alburnoides* (Steindachner, 1866)				
● *Pararutilus frisii* (Nordmann, 1840)			M	234
● *Pararutilus frisii meidingeri* (Heckel, 1852)	+	R–E	M	233
● *Scardinius erythrophthalmus* (Linnaeus, 1758)	+	V	M	236
● *Scardinius graecus* Stephanidis, 1937			M	237
● *Scardinius racovitzai* G. Müller, 1958			M	237
● *Tinca tinca* (Linnaeus, 1758)	+	I	M	240
● *Vimba vimba* (Linnaeus, 1758)	+	I–R	M	243

Cobitidae

● *Cobitis taenia* Linnaeus, 1758	+	R	M	247
● *Cobitis elongata* Heckel & Kner, 1858				
● *Cobitis (Sabanejewia) aurata* (Filippi, 1865)	+	R–V	M	250
● *Cobitis trichonica* Stephanidis, 1974				
● *Cobitis (S.) caspia* Eichwald, 1838				
● *Cobitis (S.) caucasica* Berg, 1906				
● *Cobitis (S.) conspersa* Cantoni, 1882				
● *Cobitis (S.) larvata* Filippi, 1859				
● *Cobitis (S.) romanica* Bacescu, 1943				
● *Misgurnus fossilis* (Linnaeus, 1758)	+	R–V	M	253
● *Noemacheilus barbatulus* (Linnaeus, 1758)	+	R–V	M	256
● *Noemacheilus angorae* Steindachner, 1897				
● *Noemacheilus merga* (Krynicki, 1840)				

Ictaluridae (Ameiuridae)

○ *Ictalurus melas* (Rafinesque, 1820)				
○ *Ictalurus nebulosus* (Le Seur, 1819)				
○ *Ictalurus punctatus* (Rafinesque, 1818)				

Siluridae

● *Silurus glanis* Linnaeus, 1758	+	R–V	M	259
● *Silurus (Parasilurus) aristotelis* Garman, 1890				

Gadidae

● *Lota lota* (Linnaeus, 1758)	+	R–I	M	263

Cyprinodontidae

● *Aphanius fasciatus* Cuvier & Valenciennes, 1821				
● *Aphanius iberus* Cuvier & Valenciennes, 1846				
● *Valencia hispanica* (Cuvier & Valenciennes, 1846)				

Poecilidae

○ *Poecilia reticulata* Peters, 1859				
○ *Gambusia affinis* Baird & Girard, 1853				

Atherinidae

● *Atherina (Hepsetia) hepsetus* Linnaeus, 1758				
● *Atherina (Hepsetia) boyeri* Risso, 1810				
● *Atherina (Atherina) presbyter* Cuvier, 1829				

Syngnathidae

● *Syngnathus nigrolineatus* Eichwald, 1831				

Gasterosteidae

● *Gasterosteus aculeatus* Linnaeus, 1758	+	I–R–V	M	267
● *Pungitius pungitius* Linnaeus, 1758	+	I–R–V	M	271
● *Pungitius platygaster* (Kessler, 1859)				

Cottidae

● *Cottus gobio* Linnaeus, 1758	+	V	M	274
● *Cottus ferruginosus* Heckel & Kner, 1858				
● *Cottus poecilopus* Heckel, 1836	+	R–V	M	278
● *Cottus petiti* Bacescu & Bacescu, 1964				
● *Myoxocephalus quadricornis* (Linnaeus, 1758)				

Serranidae

● *Roccus labrax* (Linnaeus, 1758)

Centrarchidae

○ *Ambloplites rupestris* (Lacépède, 1802)
○ *Lepomis gibbosus* (Linnaeus, 1758)
○ *Lepomis cyanellus* Rafinesque, 1819
○ *Micropterus dolomieu* Lacépède, 1802
○ *Micropterus salmoides* (Lacépède, 1802)

Percidae

● *Perca fluviatilis* Linnaeus, 1758				
● *Gymnocephalus (Acerina) acerina* (Güldenstädt, 1775)	+	V	M	280
● *Gymnocephalus baloni* Holčík & Hensel, 1974	+	V	M	282
● *Gymnocephalus (Acerina) cernuus* (Linnaeus, 1758)	+	I	M	285
● *Gymnocephalus (Acerina) schraetser* (Linnaeus, 1758)	+	E	M	288
● *Percarina demidoffi demidoffi* Nordmann, 1840				
● *Romanichthys valsanicola Dumirescu,* Bănărescu & Stoica, 1957	+	E	M	291
● *Stizostedion lucioperca* (Linnaeus, 1758)	+	I–V	M	294
● *Stizostedion marinum* (Cuvier & Valenciennes, 1828)				
● *Stizostedion volgensis* (Gmelin, 1788)	+	I–V	M	297
● *Zingel zingel* (Linnaeus, 1766)	+	E	M	300
● *Zingel asper* (Linnaeus, 1758)	+	E	M	303
● *Zingel streber* (Siebold, 1863)	+	E	M	305

Cichlidae

○ *Astronotus ocellatus* (Cuvier, 1829)
○ *Cichlasoma facetum* (Jenys, 1842)

Mugilidae

● *Mugil cephalus* Linnaeus, 1758
● *Mugil laber* Cuvier, 1829
● *Mugil saliens* Risso, 1810
● *Liza aurata* (Risso, 1810)
● *Liza ramada* (Risso, 1826)
● *Chelon labrosus* (Risso, 1826)

Blennidae

● *Blennius fluviatilis* Asso, 1801				

Gobiidae

● *Gobius (Apollonia) melanostomus melanostomus* Pallas, 1811				
● *Gobius (Babka) gymnotrachelus* Kessler, 1857				
● *Gobius (Mesogobius) batrachocephalus* Pallas, 1811				
● *Gobius (Mesogobius) nonultimus* Iljin, 1936				
● *Gobius (Neogobius) fluviatilis* fluviatilis (Pallas, 1811)	+	R	M	308
● *Gobius (Ponticola) kessleri* Günther, 1861				
● *Gobius (Ponticola) syrman* Nordmann, 1840				
● *Gobius (Zosterisessor) ophiocephalus* Pallas, 1811				
● *Benthophiloides brauneri* Beling & Iljin, 1927				
● *Benthophilus macrocephalus magistri* Iljin, 1927				
● *Benthophilus stellatus stellatus* (Sauvage, 1874)				
● *Padogobius panizzai* (Verga, 1841)				
● *Pomatoschistus canestrini* (Ninni, 1882)				
● *Pomatoschistus microps* (Kröyer, 1838)				
● *Proterorhinus marmoratus* (Pallas, 1811)	+	V	M	311

Channidae

○ *Ophicephalus (Ophicephalus) argus warpachowski* (Berg, 1909)				

Pleuronectidae

● *Platichthys flesus* (Linnaeus, 1758)	+	R–V		314
● *Pleuronectes (Liopsetta) glacialis* (Pallas, 1776)	+	I		315

3. Systematic part

3.1. How much are selected species endangered

The classification system adopted here follows the model published by IUCN in 1974, in accordance with the Council of Europe's wish to retain this internationally established scheme. Needless to say, it may be considered suitable better for terrestrial living animals which are visible and/or countable. As mentioned earlier (LELEK, 1977), it es extremely difficult to prove absolute extinction of one particular taxon. Scattered records occasionally refer to the presence of a rare species usually to the catch of one individual, which has not been seen in a given area of its former distribution for years. Such records are more or less meaningless, e. g. a salmon from the Upper Rhine in 1978, a solitary bitterling caught in the Rhine (km 474), one among 10,000 individuals of other species caught during the last 4 years, a single sturgeon caught between the Elbe estuary and Helgoland in 1970. Species that occur very rarely and exceptionally or as single, solitary individuals without any evidence of the maintenance of a population, are considered as extinect. The meaning of the categories endangered, vulnerable, rare and indeterminate are therefore defined as follows:

Classification of status

Endangered **(E)**: All species, including those supposedly extinct, with only a "theoretical" chance of being re-established or re-introduced within the former range of their distribution area. This category includes threatened species which occur in one or two small localities, where their abundance is in addition very low.

Vulnerable **(V)**: Taxa that are rare and very sensitive to environmental changes caused by man. Such species may be common in unsuitable habitats, where their presence depends on artificial hatching, stocking and management.

Rare **(R):**	This category includes taxa which were never abundant but were permanently present in suitable biotopes, and species which are rare today, as their formerly large distribution areas are now reduced to a few remaining suitable biotopes.
Intermediate **(I):**	Taxa which are common now, but which may be transferred to the categories **V** or **R** if there is any further deterioration of their biotopes.

3.2. Petromyzontidae

Petromyzon marinus Linnaeus, 1758

E. Sea lamprey
F. Lamproie marine
G. Meerneunauge

Vernacular name: –

Family: Petromyzontidae

Classification: vulnerable

Distinguishing characteristics

Anguilliform, mouth with sucking disc, seven pairs of gill openings and a single median nostril between the eyes. – These are the simple general characteristics of all lampreys. The sea lamprey has the dorsal fin divided into an anterior and a posterior lobe with a space between them. The sucking disc is armed with many teeth arranged in radial rows. The upper part of the body and the sides have a marbled pattern. Length greater than 50 cm, can reach over 100 cm when mature.

Worldwide distribution

Common on both sides of the Atlantic Ocean. On the eastern side, from Iceland, the Faroe Islands and northern Norway, southward to northwest Africa, including parts of the Mediterranean and Adriatic Seas. On the western side, from southwest Greenland and the Grand Banks, the Gulf of St. Lawrence and as far up the St. Lawrence River and its tributaries as Sorel, Quebec and southward to northeast Florida. Rare in the freshwaters of Newfoundland. On the American continent it lives not only as an anadromous species, but also as a landlocked, non-migratory lamprey. For details see Applegate (1950) or the almost comprehensive literature of Shelley and Beamish (1977) and Vol 1/I of "The Freshwater Fishes of Europe".

Distribution in Europe

The European coasts from Scandinavia, northeast to the White Sea, and southwards from the northern coasts of the Mediterranean to the Adriatic. It is also found in the middle reaches of the rivers Po, Adige, Arno and Tiber, in the lower reaches of the rivers Ticino, Lambro, Scrivina, in several rivers in Vento, supposedly on Sicily, Sardinia and Corsica. Common in Portuguese rivers (Almaça, pers. comm.) and Cervia (pers. comm.) reported its common occurence in rivers of northern Spain including the Guadalquivir. It is rare on the southern shores of Iceland and the Faroe Islands and in the Baltic, though it has been observed as far eastwards as the Vistula and Narev rivers.

Ecology

This lamprey is anadromous and parasitic as adult. Spawning takes place only in freshwater, very far from the estuaries. For example Frič (1908) reported individuals in a tributary to the Vltava (Moldau) River in southwest Bohemia at spawning time. This is over 600 km from the estuary of the Elbe River. Eggs are laid in nests prepared by the female. The adults die after mating. Larvae and juveniles (known as ammocoetes) spend at least 4 years as sedentary filter feeders. Their summer diet consists largely of minute plant and animal material such as filamentous green algae, detritus, diatoms, desmids and protozoa. They can withstand extremely long periods of starvation. Soon after transformation (length about 20 cm), they descend to the sea. If a host is avialable, they usually attach themselves to it, low on the body between the paired fins by means of the sucking disc. The teeth of the tongue rasp a hole in the skin of the host, and its body fluids and blood are consumed. The parasitic feeding usually takes place in the sea. It preys on salmon, sea trout, cod and other fishes. The exact length of time spent in the sea is unkonwn. When mature, the lampreys migrate upstream and do not feed; the gonads develop and ripen. Maturing animals may be observed occasionally in the open sea. A sub-adult lamprey was caught 350 km off Ireland in the eastern North Atlantic (Lelek, 1973).

Population in Europe

Lampreys are not common in northern Europe, but the larvae seem abundant in those streams they now inhabit. In the rivers of Portugal, where they are common, they are caught during the winter months. Several adult individuals were caught in a tributary to the Rhine (River Mosela) recently (Pelz 1985).

Reasons for decline

It is assumed that the pollution of estuaries and the building of weirs and dams prevent the sea lamprey from reaching their spawning grounds.

Conservation measures taken

Unknown.

Conservation measures proposed

It is suggested that free access be maintained to some streams where lamprey are known to occur, particularly in northern Europe. Traditional spawning grounds in the central European rivers, including the Rhine and Elbe, are already irretrievably lost. Hopefully some streams will be maintained free from barriers for the spawning in Sweden, Finland, the British Isles and Northern Germany.

Caspiomyzon wagneri (Kessler, 1870)

E. Caspian lamprey

F. Lamproie caspienne

G. Kaspisches Neunauge

Vernacular name: Kaspijskaja minoga (Russian)

Family: Petromyzontidae

Classification: vulnerable

Distinguishing characteristics

This species is simular to the sea lamprey, but much smaller, reaching a maximum length of about 50 cm; colour brown, without a marbled pattern. For detailed informations, see Berg (1948–1962, Vol. 1) and Vol 1/I of "The Freshwater Fishes of Europe".

Distribution in Europe

This species is present at the edge of the European region: in the Caspian Sea and the Ural, Terek and Kura Rivers. It disappeared from Upper Volga after the construction of huge, man-made lakes (Podubnyi 1948).

Ecology

Anadromous species with parasitic feeding habits after metamorphosis, like the sea lamprey, but apparently with a shorter life span.

Conservation measures taken

Unknown.

Conservation measures proposed

It would be advisable to make sure that this lamprey is not cut off from its spawning grounds in the rivers flowing into the Caspian Sea. Its situation is the same as that of the sea lamprey in northwestern Europe.

Eudontomyzon danfordi Regan, 1911

E. Carpathian lamprey
F. Lamproie carpathique
G. Donauneunauge

Vernacular name: –

Family: Petromyzontidae

Classification – vulnerable

Distinguishing characteristics

Small species, maximum length 25 cm, occurring only in the Danube catchment area. For identification purposes, it is important to study the dentition of the sucking disc. The lobes of the dorsal fin are separated from each other. Recommended literature with drawings: BĂNĂRESCU (1969), LADIGES and VOGT (1965), BERG (1948–1962, vol. 1) and particularly Fig. 23 in Vol. 1/I of "The Freshwater Fishes of Europe".

Distribution in Europe

Confined to the watershed of the Tisza River in the Danube catchment area (see distribution map).

Ecology

The larvae have feeding habits similar to those of all other lampreys. According to KUX (1965), the Danube lampreys feed after their metamorphosis during the months of July and August. Then they do not take food again until March or April of the following year. The digestive tract is empty during this resting period and the gonads are scarcely developed. Beginnung in spring, they prey upon fish and/or other fleshy food organisms, including fish fry and amphibians. They stop feeding in the autumn. During the winter, the gonads develop and spawning takes place the following spring, after which the lampreys die. This species does not migrate. Its life span is from 6 to 7 years.

Reasons for decline

Danube lampreys remain abundant only as long as there are no changes in the environment. Construction projects along rivers that cause rapid

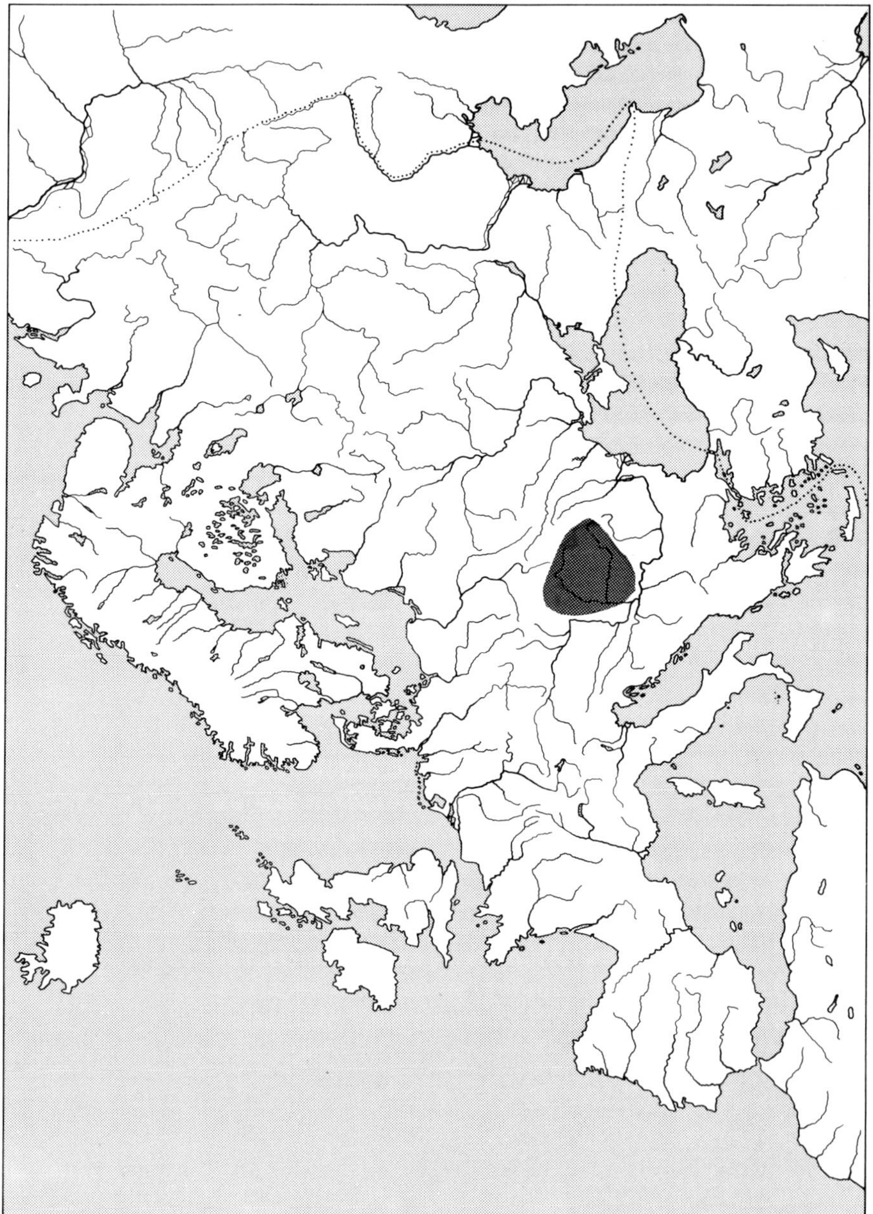

Fig. 1: Distribution of *Eudontomyzon danfordi*, Regan 1911.

changes in the water level or the disappearance of debris deposited on the bottom may have a detrimental effect on the populations.

Conservation measures proposed

Environmental protection for the conservation of lampreys is discussed in Chapter 4, p. 316.

Eudontomyzon mariae (Berg, 1931)

E. Ukrainian brook lamprey
F. Lamproie de ruisseau ukrainienne
G. Donau Bachneunauge

Vernacular name: –

Family: Petromyzontidae

Classification: vulnerable

Note: The taxonomic classification of this species remains to be defined. The validity of three subspecies is assumed: *Eudontomyzon mariae mariae, E. mariae vladykovi* (formerly described as a species), and *E. stankokaramani.*

Distingushing characteristics

This species differs from *E. danfordi* in its smaller size, the non-parasitic mode of life, the absence of wings on the velar apparatus and fewer rows of anterials. Freshly preserved larvae may be distinguished from other species of the genus by the colouration, which fades away by the time. For this reason, ammocoetes may be often confused with other species.

Distribution

The Ukrainian brook lamprey is the most widespread species of the genus. It is found in the rivers around the Adriatic, Aegean, Azov, Baltic and Black Seas (Map, see page 175, Vol. 1/I).

Ecology

E. mariae is a non-migratory, non-parasitic lamprey. It inhabits foothill zones of rivers with clear water, strong current and a gravel-sand substrate. The larvae feed on diatoms and detritus; after metamorphosis no food is taken. The life span of adults is only about 7 months.

Reasons for decline

The populations of *E. mariae* are endangered like all lamprey species in Europe, by loss of suitable habitat. They rely on clear water and are sensitive to all kind of pollution.

Conservation measures

See the text on *Lampreta planeri* p. 39, as the conservation measures are the same for all non-migratory lampreys.

Eudontomyzon hellenicus, Vladykov, Renaud, Kott and Economidis 1982

E. Greek brook lamprey
F. Lamproie de ruisseau greque
G. Griechisches Bachneunauge

Vernacular name: –

Family: Petromyzontidae

Classification: vulnerable

Distinguishing characteristics

This non-parasitic freshwater lamprey differs from the other European species of *Eudontomyzon* by its smaller size, fewer trunk myomeres, and rounded caudal fin. See Vol. 1/I of "The Freshwater Fishes of Europe".

Distribution

E. hellenicus is endemic to two rtiver systems: the Strymon in the Aegean Sea Drainage Area of Macedonia, and the Loúros River.

Ecology

Not much is known about its ecology. It lives in clear, fairly fast-flowing brooks. The adults undergo very local spawning migrations. The ammocoetes are filter feeder; the adults, non-parasitic.

Reason for decline and conservation measures

Population sizes and densities are not known. This strongly endemic species is vulnerable because of the extreme small size of its range. Extinction may be caused by regional pollution or local changes in the river systems inhabited by the Greek brook lamprey.

Lethenteron (Lampetra) japonicum (Martens, 1868)

E. Arctic lamprey
F. Lamproie arctique
G. Arktisches Neunauge

Vernacular name: –

Family: Petromyzontidae

Classification: vulnerable

Distinguishing characteristics

The Arctic lamprey differs in some features from the European satellite species, *Lethenteron kessleri* and from *Lampetra fluviatilis,* but it is often confused with them. For detailed information, see Vol. 1/I of "The Freshwater Fishes of Europe".

Worldwide distribution

L. japonicum inhabits great parts of the Circumpolar Subregion of the Holarctic Region in Asia, North America, and Europe.

Distribution in Europe

In Europe, the Arctic lamprey is confined to the coastal waters and rivers flowing into the Barents and White Sea.

Ecology

Not much information is available about the ecology of European stocks of this species. Some data, taken from Russian literature, are provided in Vol. 1/I of this series.

Reasons for decline

This species is listed here, because its European stocks are in danger due to increasing pollution, even in the Arctic regions and because the ammocoetes are captured for use as a natural bait.

Conservation measures

See text on *Lampetra planeri*, p. 39.

Lethenteron (Lampetra) zanandreai (Vladykov, 1955)

E. Lombardy brook lamprey
F. Lamproie de ruisseau de Lombardie
G. Lombardisches Neunauge

Vernacular name: –

Family: Petromyzontidae

Classification: vulnerable

Distinguishing characteristics

This species is a freshwater brook lamprey of moderate size. It differs from other lampreys, that occur in Italy in the number of trunk myomeres and the presence of posterials on the oral disc. Specification of morphological features, see Vol. 1/I of "The Freshwater Fishes of Europe".

Distribution

The Lombardy brook lamprey is endemic to a small region in northern Italy, from the boundaries of the Italian watershed to the left side of the Po River.

Ecology

L. zanandreai is a non-parasitic, non-migratory species. Its habitat is characterized by clean, cold water. The ammocoetes are filtering bottom-feeders; the adults do not feed at all.

Resons for decline and conservation measures

Nothing is known about changes in the size of its stocks. But this species surely is as vulnerable as the other purely freshwater lampreys. The Lombardy brook lamprey is especially endangered due to its small range and its need of clean, cold, and unpolluted water. No conservation measures have yet been taken.

Lampetra fluviatilis (Linnaeus, 1758)

E. Lampern – River lamprey
F. Lamproie de rivière
G. Flußneunauge

Vernacular name: –

Family: Petromyzontidae

Classification: endangered

Distinguishing characteristics

This is the second largest species, reaching a length of about 40 cm. The trunk myomeres are less than 60, the branchial region is unpigmented below the gill slits. For an illustration of the sucking disc, see Vol. 1/I of this series.

Distribution in Europe

Coastal waters and rivers in northwestern Europe, but not in Iceland. Found in Ireland and the British Isles, with the exception of the northernmost tip of Scotland (northward of Inverness). In Norway, it occurs only south of Varanger Fjord, in rivers flowing to the North and the Baltic Seas, including the Gulf of Bothnia. Two ecological forms occur in Lakes Onega and Ladoga. In southern Europe it occurs on the Iberian Peninsula, in Italy it is found in the Po River and rivers in the following regions: Lombardy, Emilia, Romagna, Vento, Toscana and in Lake di Garda. With very few exceptions it is absent from the Black Sea catchment area and Alpine lakes.

Ecology

This lamprey is anadromous and parasitic. From September to November, is ascends to the upper reaches of streams and spawning takes place from February to May. During the upstream migration, the lamprey does not eat and it dies after spawning. The larvae stay in the stream for three to four years. Metamorphosis is completed when the individuals reach a length of 14 to 15 cm. The lamprey then begins the downstream migration, feeding intensively, on fish, and growing rapidly. It stays in the sea for about two years. The early mature form, "praecox", only stays in the sea during the summer months.

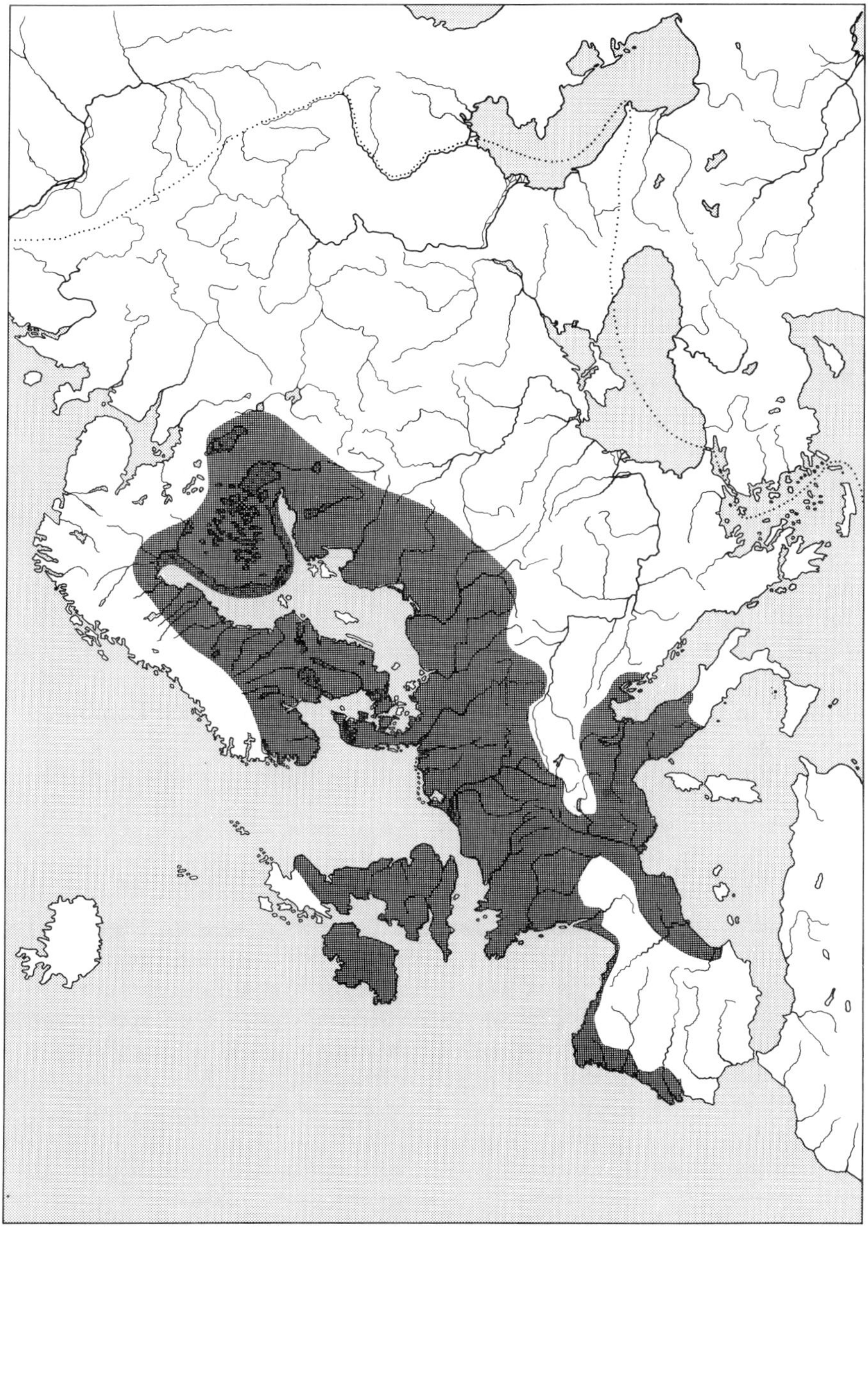

Fig. 2: European distribution of *Lampetra fluviatilis* (Linnaeus, 1758).

Population in Europe

See *Petromyzon marinus.*

Reasons for decline

The same as for *Petromyzon marinus.* The species seems to be greatly endangered, because stretches of its former habitat (lower sections of the trout zone) have been altered by weirs, the spawning sites have been eliminated, and the water quality is not good enough for such a sensitive fish.

Conservation measures taken

No effort has been made to conserve this species, though it is sold at rather high prices in the Baltic area. In the Netherlands, it is protected by law.

Conservation measures proposed

See *Petromyzon marinus.*

Lampetra planeri (Bloch, 1784)

E. Brook lamprey
F. Petite lamproie
G. Bachneunauge

Vernacular name: –

Family: Petromyzontidae

Classification: vulnerable

Distinguishing characteristics

The main distinguishing feature is again the sucking disc. For an illustration see Vol. 1/I of this series. All the teeth are weak and fairly blunt. The caudal fin has no pigment cells, unlike *Eudontomyzon danfordi*. When pigment cells are present, they occur only near the junction of the fin and the body. There is no space between the dorsal fins. Adults are from 11 to 15 cm long.

Distribution in Europe

Throughout the Baltic and the North Sea catchment areas and throughout the British Isles including Ireland, with the exception of northern Scotland. In the south: France, all of Italy, Sardinia, Sicily, Lake Garda, Albania, and probably Spain and Portugal. In the East, the uppermost reaches of the Volga. See map.

Ecology

The Brook lamprey occurred in almost all small streams within its distribution area. It regularly occupied water courses at medium altitudes where brown trouts are found. As this species remains in the stream sediment most of the time, at least during the daytime, the structure of the river bed is of primary importance. The most suitable habitats are stretches of small streams with moderately fluctuating water depths and currents, in which the riffles and pools alternate frequently.

The seasonal water discharge frequently superimposes sand on the silt layers which contain leaves and other plant material. The actual habitats required by juvenile lampreys, particularly the brook lamprey, are the transition zones between sand deposits and areas of silting.

After hatching at the spawning sites on gravel substrata, the larvae move into the habitats described above and stay there for at least 4 years, until

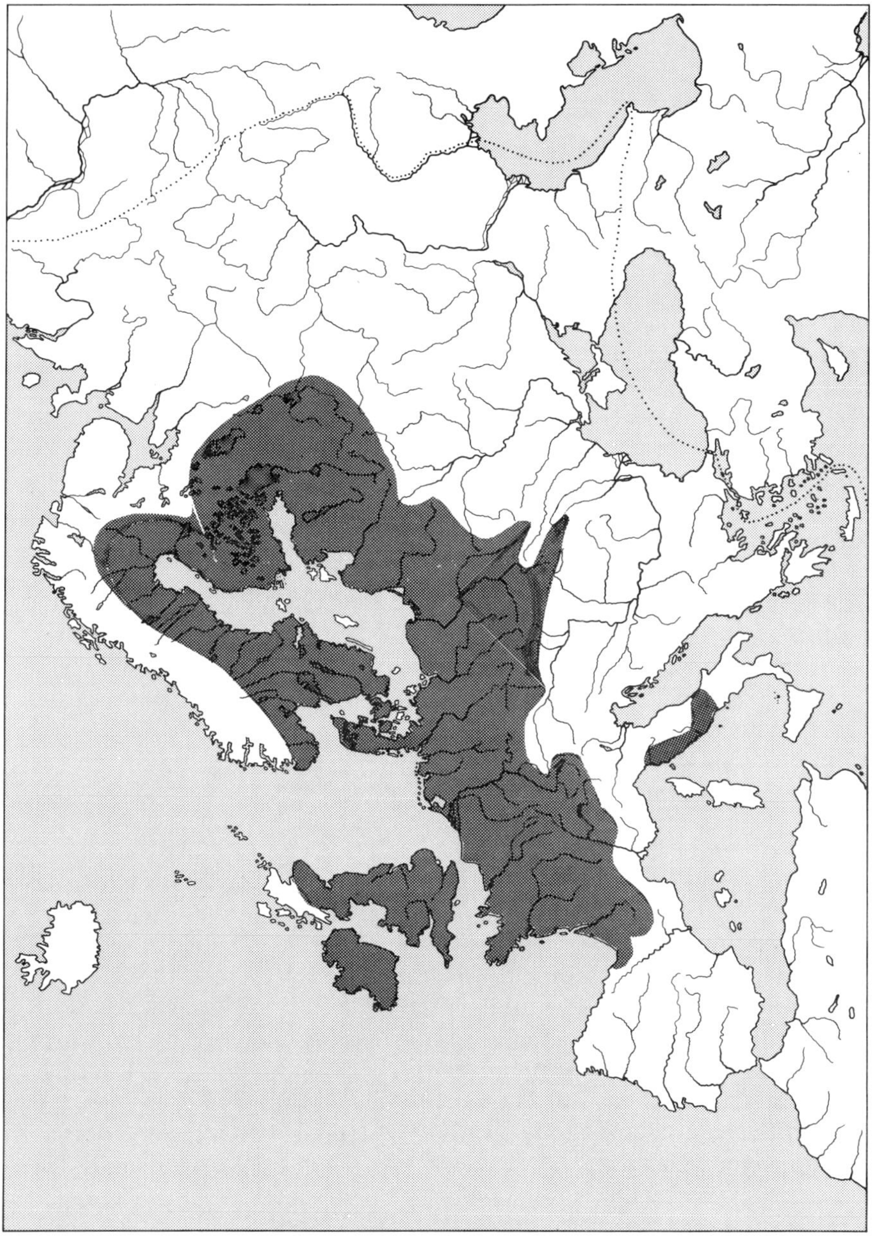

Fig. 3: Distribution in Europe of *Lampetra planeri* (Bloch, 1784).

metamorphosis is completed. Adults do not take food at all as the digestive tract no longer functions. They spawn in the early spring (March to April) and die shortly afterwards.

Population in Europe

Formerly very abundant throughout the montane area (Mittelgebirge) but now confined either to higher altitudes, where the hydrological conditions are less suitable, or to the wooded areas, where little or no modifications have been made in the water courses. The former large, more or less continuous distribution area has broken down into small islets, characterized by good water quality and the absence of man-made alterations to either the stream itself or its surroundings.

Reasons for decline

Changes in water courses, particularly artificial straightenings and construction projects, drastic seasonal variations in water quantity, and general decrease in water quality, are responsible for the local disappearance of this species. Intensive use of electrical devices for stunning fish may be detrimental to the lampreys in the long run.

Conservation measures taken

The establishment of protected areas for human water supply, in which the streams do not drain into one central reservoir, has saved many populations. Other measures, such as a ban on collecting, are of little importance and have slight effect. In the Netherlands, the species is protected by law.

Conservation measures proposed

Conservation of the suitable environment is the only solution to the problem of protecting this species. There would then be no danger of its extinction, as the amount of living space required is small and there is no need to provide special spawning or feeding grounds. The life-span, though rather long, can be passed entirely in a relatively small stretch of stream.

3.3. Acipenseridae

Acipenser sturio Linnaeus, 1758

E. Sturgeon
F. Esturgeon
G. Stör

Vernacular name: –

Family: Acipenseridae

Classification: endangered (close to extinction)

Distinguishing characteristics

All sturgeons have elongated snouts, usually five rows of bony scutes along the body and the heteroceral tail fin.

The most obvious features are the five rows of large, hard, bony scutes along the body. The first row along the centre of the back has 10 to 15 plates, paired behind the dorsal fin. The lateral rows have 26 to 36 plates. On the belly, there are two rows of 9 to 13 plates, which merge into a paired central row behind the vent. The snout is long, with two pairs of barbels without ramification, midway between the tip and the mouth. Sturgeon may grow to over 3 m; females are larger than males.

Worldwide distribution

Sturgeons are confined to the northern hemisphere on both sides of the Atlantic and Pacific.

Distribution in Europe

The coastline of Europe from the North Cape through the Mediterranean to the Black Sea. (See distribution map on page 43, which gives past and possible present distribution.)

Ecology

The sturgeon is a rare visitor to northern European waters. Once moderately abundant in coastal waters, it has now been practically exterminated. The adult sturgeons enter the river mouth in early spring, the males being

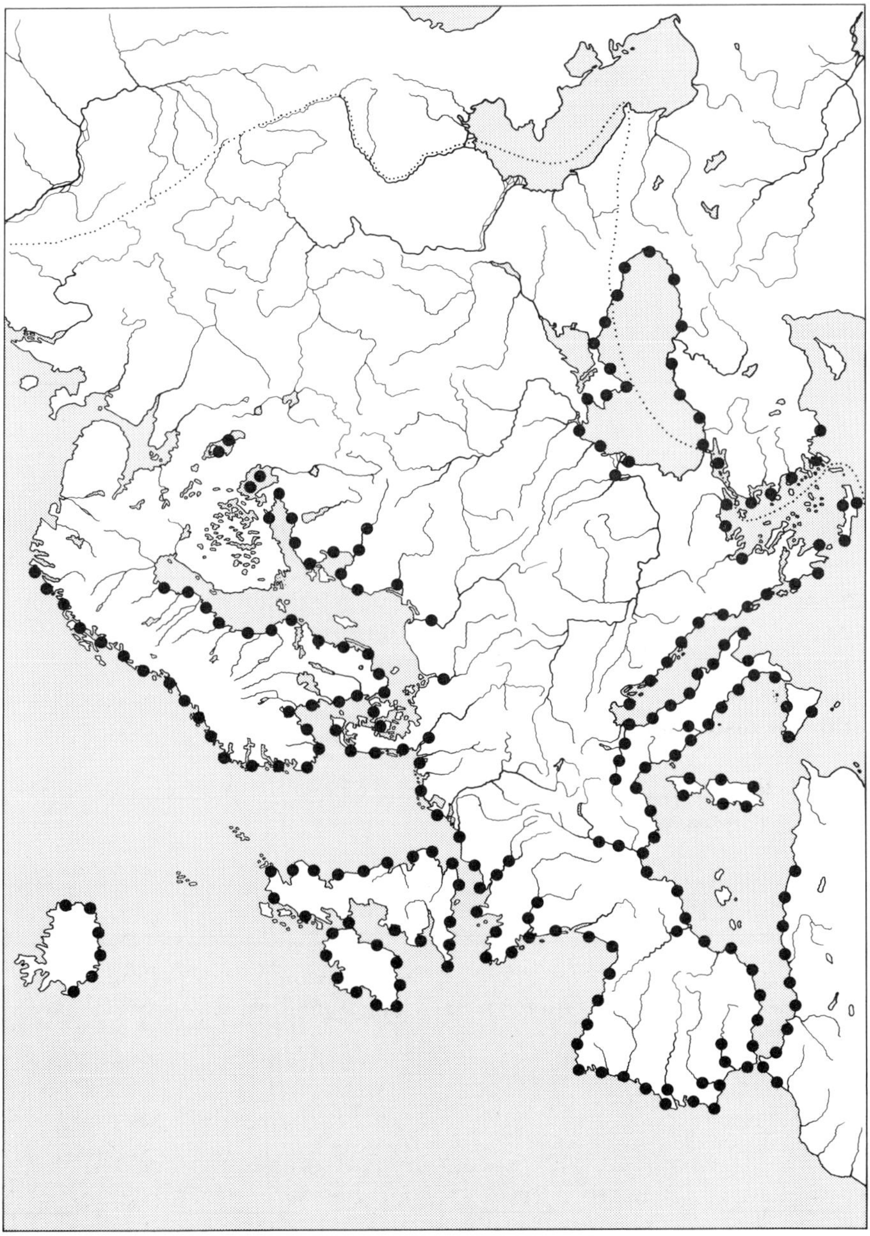

Fig. 4: European distribution of *Acipenser sturio* Linnaeus, 1758.

more numerous than the females. Spawning takes place in deep (about 6 to 8 m) swiftly flowing water with gravel bottom. According to BALON (1975), sturgeons are litho-pelagic spawners. After releasing the eggs, the adults return immediately to the sea, but the young stay in the river or its estuary for 2 or 3 years. Males reach maturity between 7 and 9 years of age, females from 8 to 14. The newly mature fish are already from 1 to 1,5 m long. Juveniles feed mainly on insect larvae, isopods, gammarids, molluscs and small fish. Adults eat molluscs, small fish (gobies and sand eels), polychaete worms, shrimps and isopods. They probe the bottom mud and sand with their snouts, detect edible items with the sensitive barbels and pick them up with their protrusive lips.

Population in Europe

Only a very few populations still exist in Europe: in the area near the mouth of the Gironde (France), the Guadalquivir (Spain), and the Danube. A freshwater population inhabits further Lake Ladoga (USSR). Detailed informations on populations in the USSR will be presented in Vol. 1/II of "The Freshwater Fishes of Europe".

It has been seen in British rivers and estuaries, but it has never been proved that it spawns there. Formerly, the sturgeon was found as far upstream as Basel (Rhine), Krakow (Vistula), Würzburg (Main) and Prague (Vltava–Moldau), but this is no longer the case. For details, see DUNCKER (1960), MOHR (1952), and VERHEY (1949). Its former frequent occurrence in the Baltic is reviewed by DUNCKER (1960).

Reasons for decline

Its spawning grounds have been altered or damaged by construction works to make the river navigable, or weirs and locks have made access impossible, as in the Guadalquivir River (Cervia, pers. comm.). Moreover, overfishing in the past siltation and poor water quality in the present have brought this species near to extinction in northern Europe.

Conservation measures taken

Up to the present time, only closed seasons, size limits and catch limits have guaranteed the survival of the remaining European populations. Environmental protection measures are not known.

Conservation measures proposed

It is hoped to achieve some recovery by improving the water quality in estuaries. This measure, if feasible, may help newly hatched young to survive. As a substitute for migration to the spawning grounds (mostly already eliminated), induced spawning as practised by some Black and Caspian Sea sturgeon fisheries might be tried, followed by stocking the lower reaches of large rivers and estuaries, but existing protective measures must be maintained.

Acipenser güldenstaedti Brandt, 1833

E. –

F. –

G. Waxdick

Vernacular name: Osetr (Russian)

Family: Acipenseridae

Classification: vulnerable

Subspecies

Acipenser g. güldenstaedti (Brandt, 1833) (northern Caspian Sea)
Acipenser g. persicus (Borodin, 1897) (southern Caspian Sea)
Acipenser g. colchicus (Marti, 1940) (Black Sea area)

Distinguishing characteristics

For general characteristics, see *A. sturio*. This species possesses 10 to 13 separated dorsal scutes, 21 to 50 rhomboid lateral scutes and 8 to 10 ventral scutes. Barbels are located close to the tip of the snout. They are short and round, without fringes. Rather a large fish, over 2 m in length.

Distribution in Europe

This species is found only in the Black and Caspian Seas. It is an anadromous fish, ascending the Danube as far as just beyond Bratislava. A 17th century record mentions Regensburg (Balon 1967). Occasionally in the Gulf of Finland as a result of USSR stocking.

Ecology

It is similar to that of other species of the Acipenseridae. There is a stable freshwater population in the Volga Delta. Spawning need not necessarily be on gravel bottoms but can take place on the sandy stretches at the mouths of tributaries as well. The larvae and fry are negativly phototactic.

Reasons for decline

If fishing increases, the stock may well become endangered, as like other sturgeons, it reaches maturity relatively late, at 10 years old or more.

Conservation measures taken

Catches are limited, and there are regulations on size and a closed season. In addition, artificial propagation and stocking are carried out successfully.

Conservation measures proposed

See *A. sturio.*

Acipenser naccarii Bonaparte, 1834–1841

E. – Vernacular name: Storione cobice (Italian), Esturgião (Portuguese)
F. –
G. – Family: Acipenseridae

Classification: vulnerable

Distinguishing characteristics

Possesses all the general features of sturgeons. Typical meristic patterns: 12 to 14 dorsal scutes, 40 to 42 lateral scutes. The skin is densely covered with large and small bony structures. The snout is blunt and about one-third of the head length. The barbels are unramified, rather long, and located closer to the snout than to the eyes.

Distribution

Adriatic Sea, Po and Adige Rivers, rivers of the provinces of Lombardy and Veneto.

Ecology

Anadromous species, its ecology is not known in detail.

Conservation measures

Very little is known about this species, and before any conservation measures can be taken, additional information is desperately needed.

Acipenser nudiventris Lovetsky, 1828

E. –

F. –

G. Glattdick

Vernacular name: Ship (Russian)

Family: Acipenseridae

Classification: vulnerable to endangered

Distinguishing characteristics

A. nudiventris has the typical sturgeon habitus with 12 to 15 dorsal scutes and over 60 lateral scutes. The 12 to 16 scutes on the underside are sometimes absent. Unlike other sturgeons, the lower lip is not divided. Barbels are rounded with ramifications, those on the left hand side of the snout pointing right, those on the right pointing left. The body is conspicuously stout and compact.

Distribution

This sturgeon species lives in the western parts of the Black and Caspian Seas and was introduced into the Aral Sea. It is a typical anadromous sturgeon; it may stay in freshwater during the winter months, if the spawning run takes place in the autumn. It has been seen in the Danube as far upstream as Bratislava (Slovakia), river km 1869. The strongest run was observed in the Ural River.

Ecology

It is comparable to that of other sturgeons, but this species may be more vulnerable to environmental changes and overfishing. Furthermore, it reaches sexual maturity late and does not spawn every year, but sometimes only every second or third year. Construction projects and giant man-made lakes on the rivers leading to the Caspian Sea prevent its migrations upstream. Nikolskij (1957) already believed it to be endangered. One reason is that the fish has to spend a long period in freshwater to stimulate the ripening of the eggs and sperm.

Population in Europe

The population is rather small and is confined to two river basins.

Conservation measures taken

Unknown.

Conservation measures proposed

Artificial propagation, including induced spawning and stocking, may be useful.

Acipenser persicus Borodin, 1897

E. – Vernacular name: –
F. –
G. – Family: Acipenseridae

Classification: endangered

Note: The two species of sturgeons *A. güldenstaedti* and *A. persicus* are extremely difficult to distinguish. This taxonomic difficulty was overcome by means of serological examinations. *A. persicus* shows a basic difference from *A. güldenstaedti,* the Russian sturgeon, in the antigenetic components of its blood serum proteins. No differences were found in osteological characters. However, several ecological characteristics speak for the validity of the species. Detailed account of information and references are provided in Vol. 1/II of this series.

Distribution

A. persicus is found in the Caspian Watershed. It enters mainly the Kura, Volga and Ural Rivers for spawning. Another population migrates into the Terek, Sulak, and Samur and other rivers. It is possible that *A. persicus* also inhabits the Black Sea.

Ecology

This species prefers warmer water than the Russian sturgeon, *A. güldenstaedti.* During migration into the rivers, which may take place for several weeks, they can usually be found in the main channel of the river, near to the bottom. Sexual maturity is reached in the Kura River at 8 (males) and 12 (females) years, in the Volga it occurs about five years later. *A. persicus* may reach an age of more then 40 years (maximum age was 48). In length and weight it can exceed *A. güldenstaedti.* The average length of males in the Volga is 151 cm, and that of females is 156 cm. Females are usually larger than males of the same age. The growth rate is strongly dependant on the food supply. The diet changes during the life and resembles to that of *Huso huso*.

Reasons for decline and conservation measures

See *Acipenser sturio* and *A. güldenstaedti.*

Acipenser ruthenus Linnaeus, 1758

E. –
F. –
G. Sterlet

Vernacular name: Sterljad

Family: Acipenseridae

Classification: endangered

Distinguishing characteristics

This species has all the general features of the sturgeon family (see *Acipenser sturio*). Arrangement of scutes: Dorsal 12 or 13 to 14 or 16. The lateral row has (58) 64 to 66 (71), the underside 12 to 16. Other meristic features, according to Oliva and Hrabě (1973): D – 39 to 49; A – 20 to 30. The barbels are long with ramification. Colouring is conspicuous, upper side, dark brown or grey, protruding scutes, ivory or cream; belly, light yellow or slightly pinkish-white. Individuals from the Danube are smaller (about 50 to 60 cm long) than those from the Volga, which may reach lengths of over 100 cm.

Distribution in Europe

The Sterlet is found in freshwaters of the catchment areas of the Black and the Caspian Seas in the East, in the catchment area of the Dvina in the North and farther eastwards to the rivers Ob, Kolyma and Yenisei. Its occurrence in northern Europe was always low, at present its natural occurrence is even uncertain. In central and eastern Europe it is found in the Danube and the Volga. It can tolerate salt water, as in the north of the Caspian Sea. In the Danube catchment area, it occurs only in larger tributaries. Its present distribution upstream from Hungary only includes the Danube itself and the occurrence in the upper Danube (Austria and Germany) depends on stocking.

Ecology

Although little is known about the life cycle of this species, it is certain that it makes some spawning migrations. It moves in schools upstream to places with gravel and stony substrate and with swift current in the spring. There it spawns, supposedly in the same way as *Acipenser sturio*. Winter is spent in deep places near the river bottom, far downstream from the spawning places.

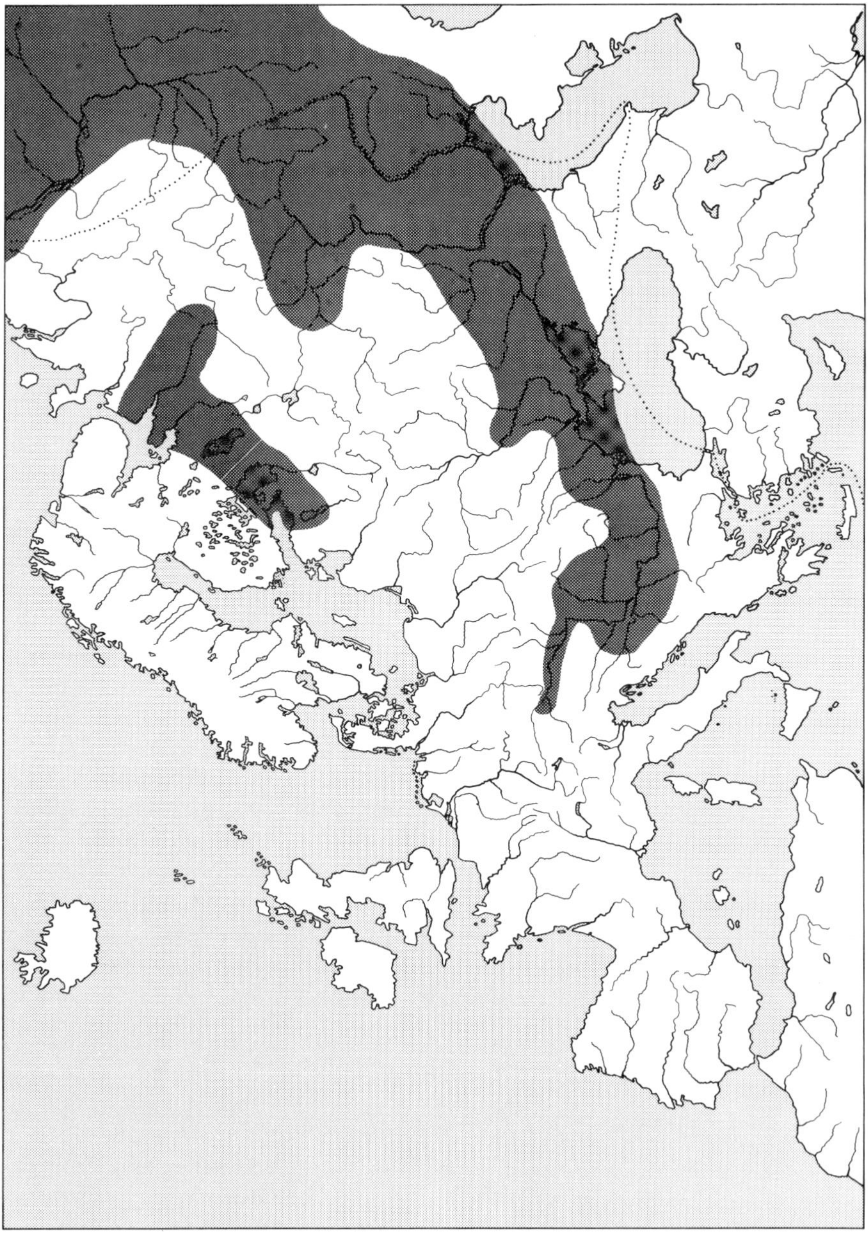

Fig. 5: Distribution in Europe of *Acipenser ruthenus* Linnaeus, 1758.

Males mature at the age of 4 to 5 years and females at 5 to 9. The fry of this species makes short daily migrations, at night towards the shore and during the daytime to deeper sections of the stream. They feed on insect larvae and drifting insects, which they snap up from the water's surface while swimming upside-down.

Reasons for decline

Pressures from human civilization were responsible for the decline: overfishing since historical times and the recent worsening of the water quality had the greatest effect, while navigation and engineering works along the river and on the banks did not affect the sterlet as much as many other species. For further details, see *Huso huso*.

Conservation measures taken

These include the imposition of size limits and closed seasons. Active measures include an attempt to introduce young sterlets into the impounded waters of man-made lakes in the Austrian and German stretches of the Danube, but the results are not known (Balon 1967, p. 59). The stocking done in last years met with success.

Conservation measures proposed

See *Acipenser sturio*.

Acipenser stellatus Pallas, 1771

E. –
F. –
G. Sternhausen

Vernacular name: Sevrjuga (Russian)

Family: Acipenseridae

Classification: vulnerable

Distinguishing characteristics

Typical sturgeons, with the following meristic patterns: 12 to 16 dorsal scutes, 30 to 40 rhomboid lateral scutes. The snout is very long: 62 to 65 % of the length of the head. At first glance, this species looks very slim.

Distribution in Europe

The main distribution areas are the Black, Caspian and, according to Tortonese (1956), probably the Adriatic Sea. It is an anadromous species usually ascending the Volga, Ural and Danube River, and some of their tributaries, and into streams flowing to the Black Sea from the north.

Population in Europe

It seems rather unstable, particularly in the Adriatic. In the easternmost parts of Europe, particularly in the Volga Delta this species is the most common one. In the Danube, it was always the rarest species, but it was found in this river as far upstream as Slovakia.

Ecology

The ecology is comparable with that of other sturgeons, but this species seems less bound to fresh water, except during its reproductive period.

Conservation measures

See *A. güldenstaedti*.

Huso huso (Linnaeus, 1758)

E. –

Vernacular name: Beluga (Russian)

F. –

G. Hausen

Family: Acipenseridae

Classification: vulnerable to endangered

Subspecies: Three subspecies have been described in literature, but validity is doubtful (see Vol. 1/II of this series):
Huso h. orientalis (Salnikov and Malyatskij, 1934) (eastern Black Sea)
Huso h. maeoticus (Salnikov and Malyatskij, 1934) (Azov Sea)
Huso h. caspicus (Babuschkin, 1942) (Caspian Sea)

Distinguishing characteristics

Huso huso has all the characteristics of the typical sturgeon. Meristic data: dorsal scutes 11 to 15; lateral, 40 to 60; ventral, 9 to 12. Colouring: dorsum mostly black, belly white, both colours distinctly separated. It may be the largest of all sturgeons.

Distribution in Europe

Caspian, Black and Adriatic Seas and large rivers feeding them, such as the Volga, Ural, Terek, Kura, Danube, Dniester, Desna, Bug and very probably the Po. In the Danube, it has been observed as far upstream as Bratislava and supposedly at the mouth of the March river, the left tributary to the Danube above Bratislava. The distribution map reflects its occurrence some 25 to 30 years ago, at present is this species absent or extremely rare upstream the territory of Bulgaria.

Ecology

Anadromous species with two spawning migrations, one in autumn and one in early spring. The autumn migrators stay in freshwater throughout the winter, already at the spawning places, far upstream; the spring migrators spawn closer to the estuary. The spawning ground usually has a gravel bottom to which the eggs stick. After hatching, the larvae drift

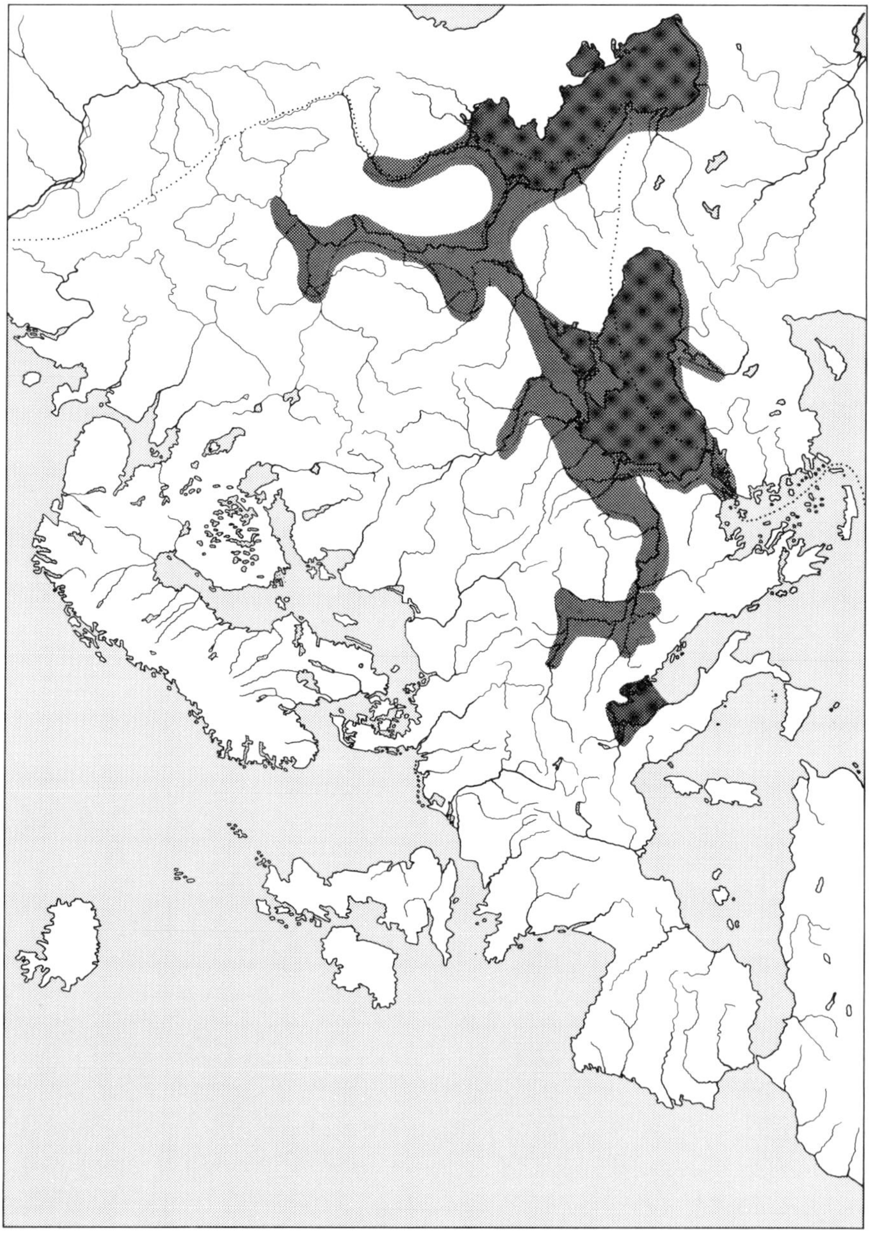

Fig. 6: Distribution of *Huso huso* (Linnaeus, 1758).

downstream, feeding on small invertebrates. Later on, in the sea, the adult feed mainly on small fish. This species is the fastest growing; it may reach 1 m in length by the fourth year of its life. But as all other sturgeons, maturity is reached very late, at 12 to 18 years.

Population in Europe

The European population has decreased constantly since the Middle Ages according to Balon (1967), who analyzed changes in the Danube stock. If all the proposed Danube dams are completed, irreparable damage to the stock must be expected, even in the delta, where this species is still intensively fished. There is a relatively large stock of this species in the Volga Delta.

Reasons for decline

Overfishing and modifications of the rivers.

Conservation measures taken

It is hoped that induced spawning and care of the freshly hatched larvae and young fish may help save this species from extinction. The best work in artificial propagation has been done in the USSR, where this species and other sturgeons are highly prized commercially for their eggs (caviar).

Conservation measures proposed

Reduction of intensity of fishing for this species, setting of new quotas for numbers to be caught, extension or introduction of closed seasons and, last but not least, promotion of more artificial propagation.

3.4. Clupeidae

Alosa alosa alosa (Linnaeus, 1758)

E. Allis shad
F. Grande alose
G. Maifisch

Vernacular name: –

Family: Clupeidae

Classification: endangered

Note: Two sub-species are important in the area under consideration:
Alosa a. bulgarica (Drensky, 1934) found in the southern part of the Black Sea and which ascends the rivers in the east of Bulgaria and
Alosa a. macedonica (Vinciguerra, 1921)

Distinguishing characteristics

Slender body, distinctly compressed laterally, eyes with conspicuous adipose eyelids; gill rakers long and numerous; scales weakly confined to body, cycloid, keel on the underside, lateral line absent. This applies to all *Alosa* species mentioned here. Meristic patterns: D 19 to 21 soft rays, A 21 to 27; in the main longitudinal row more than 70 lateral scales. Always more than 60 gill rakers on the first gill arch (90 to 130). Behind the upper margin of the opercular bone, there is always one dark spot and sometimes one or two more behind the first one.

Worldwide Distribution

The family Clupeidae, (primarily marine fishes) is found throughout the world, both in the sea and in freshwaters. In European freshwaters only the anadromous forms are important. 4 species of 2 genera are known in the North American Atlantic waters, while in Europe (if the genus *Caspialosa* is considered as valid) there are 2 genera with about 19 species and sub-species.

Distribution in Europe

Alosa alosa occurs from Norway along the coast as far south as the Mediterranean Sea. In Ireland and the British Isles, it is limited to a few localities

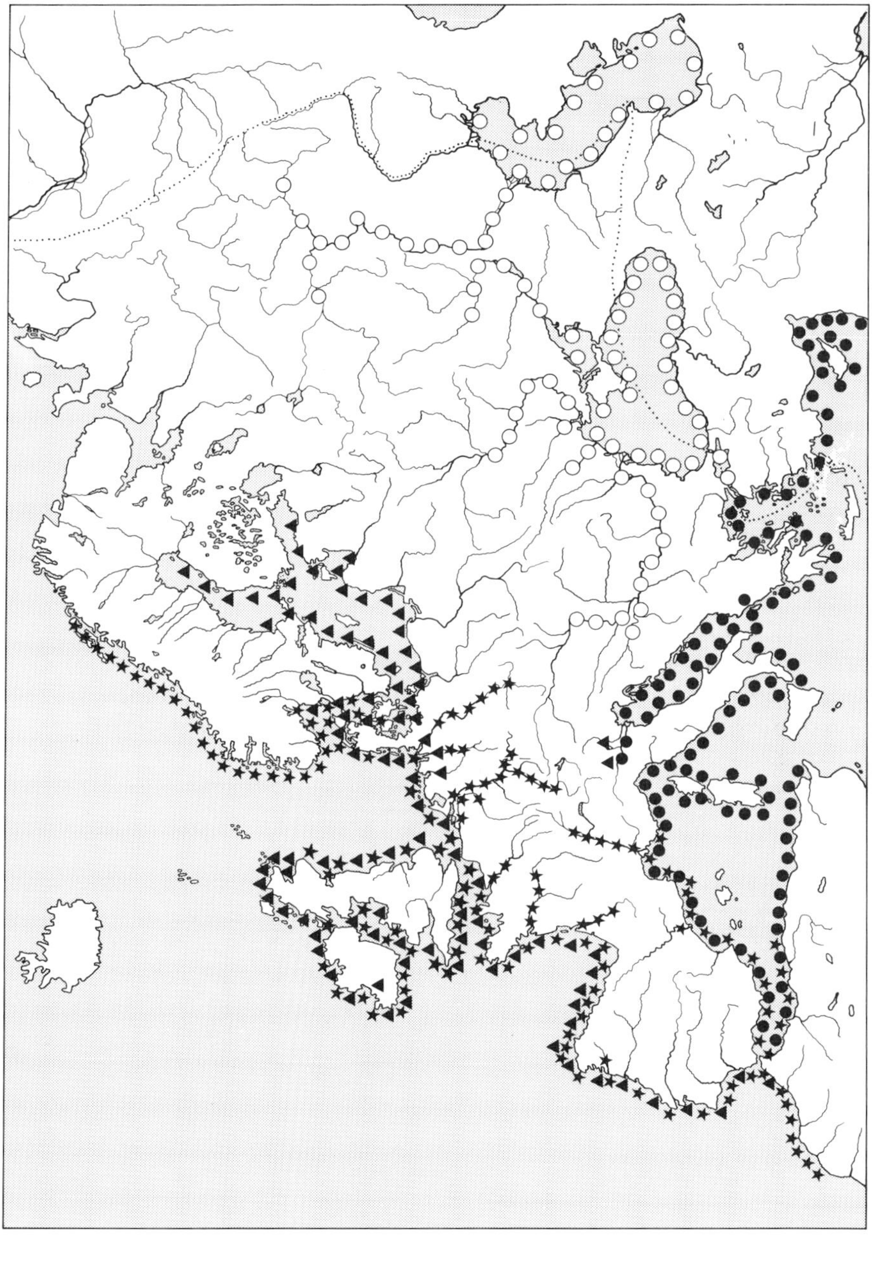

Fig. 7: European distribution of *Alosa alosa* and of *Alosa fallax*, with subspecies: ★ *Alosa a. alosa*, ▲ *Alosa f. fallax*, ● *Alosa f. nilotica*, and ○ *Alosa pontica*.

and fairly rare, occuring in estuaries and the lower reaches of rivers. It is absent from northern Scotland. Along the Atlantic coast, it enters the rivers Adoure, Garonne, Loire and Seine, but is rare in the Rhône. Formerly it migrated up the Rhine to Basel, and into its tributary, the Main. Frič (1908) recorded it from the Elbe and its tributary, the Vltava in Bohemia, and in the Rhône almost as far as Dijon (until 1952). For distribution in general, see the map on page 59.

Ecology

A typical anadromous species; spawning takes place only in freshwater, from May to June, and the eggs drift downstream. After reaching a length of 8 to 12 cm, the young fish move to the sea where they live in the pelagic zone. At 30 or 40 cm they enter freshwaters, to spawn. Individuals of 6 to 7 years old are about 70 cm in length. Until the beginning of the 20th century, the shad was an economically important species.

Population in Europe

In northern Europe, this species is almost extinct. At the mouth of the Rhine it is seen only occassionally, and in English waters it is very rare. In southern Europe it has recently become very rare (Rameye *et al.* 1976).

Reasons for decline

Poor water quality is probably chiefly responsible, and in addition the building of weirs prevents this species from reaching its spawning grounds. As proof that water quality may be decisive for its survival, it may be noted that after the Second World War, more individuals suddenly reached the markets again (Bauch 1955). Water quality had improved almost automatically in west and central Europe, as industrial plants had been in most cases destroyed resulting in less water pollution. The harmful influence of damming is clearly demonstrated on the Rhône (Rameye *et al.* 1976).

Conservation measures taken

No effective conservation measures have been taken other than legal protection.

Conservation measures proposed

Artificial propagation may help to save the species from complete extinction, and unblocked passages should be maintained between the sea and spawning places in selected river systems.

Alosa fallax fallax Lacépède, 1800

E. Twaite shad
F. Alose feinte
G. Finte

Vernacular name: –

Family: Clupeidae

Classification: endangered

Note: The subspecies important for the European continent are:

Alosa f. nilotica (Geoffroy, 1827)
French: Alose du Rhône, Italian: Cheppia. This anadromous subspecies occurs along the Mediterranean coast; its life-cycle corresponds with that of *Alosa f. fallax*. A landlocked population occurs in Lake Skuta, Albania.

Alosa f. killarnensis (Regan, 1916)
English: Killarney shad; Vernacular: Gobhar (Gaelic)
A landlocked population of this subspecies occurs in the lakes of Killarney (Ireland). It is a small plankton feeder, distinguished from *Alosa f. fallax* by more gill rakers (43 to 53).

Alosa f. lacustris (Fatio, 1890)
German: Alse; Italian: Agone
A lacustrine form inhabiting Lakes Maggiore, Lugano, Como, Iseo and Garda. Introduced into Lake Lazio.

Distinguishing characteristics.

In general, similar to *Alosa alosa.* Distinguishing features: D 18–20; A 20–27 soft rays; in the main longitudinal row, less than 70 scales; on the first gill arch usually 35 to 45 rakers but always less than 60. A row of dark spots (6 to 8) behind the operculum along the upper part of the body. This species is generally smaller than *Alosa alosa,* reaching a maximum of 55 cm.

Distribution in Europe

Along the coast of the Atlantic and the Baltic Sea, from Gibraltar to the estuary of the Narva River in the Gulf of Finland, and occasionally near to the Finnish coast including the British Isles (with Ireland). Along the northern shoreline of the Baltic, its distribution ends of the latitude of Stockholm. It is not present off the Norwegian coast, but it is found in the Bay of Oslo and to the east along the northern coastline of the North and Baltic Seas. (See map on page 59.)

Ecology

This is an anadromous species, entering the river mouths and estuaries. Spawning takes place in or just above the tidal reaches from May to June. The eggs are demersal; they drift over the estuary bottom towards the sea. Food preferences have not been much studied, but this species is assumed to take plankton. In the sea, the young shad prey on the fry of other fish.

Population in Europe

Though little is known of the population in the sea, its occurrence has decreased drastically in the last few years. At the mouths of northern European rivers, it is considered as almost extinct.

Reasons for decline

It is assumed that pollution and heavy ship-traffic are the main reasons (see *Alosa alosa*).

Conservation measures taken

Not known.

Conservation measures proposed

See *Alosa alosa*.

3.5. Anguillidae

Anguilla anguilla (Linnaeus, 1758)

E. Eel
F. Anguille
G. Aal

Vernacular name: –

Family: Anguillidae

Classification: intermediate and/or vulnerable (dependent on stocking)

Distinguishing characteristics

Though the expression "eel-like" is generally known, the zoological description is nevertheless given here: body elongated, some bony elements fused, jaw developed and toothed. Fins: all soft-rayed, dorsal long, and joined with the anal at the extremity of the caudal. Scales small and embedded. For more information on the 19 species of the genus *Anguilla,* distributed throughout the world, see Deelder (1970) or Tesch (1972). There are only two species which are difficult to distinguish and which will be considered here: the European and the American Eel, the difference between them being only the number of vertebrae, *Anguilla anguilla* having 110 to 119 and *Anguilla rostrata* 103 to 111.

Worldwide distribution

From the same spawning grounds, supposedly in the Sargasso Sea (20° to 35° N and 60° to 70° W) come the two very similar species which occupy the east of the North American continent and a large part of Europe.

Distribution in Europe

It enters the estuaries of all rivers which empty into the Atlantic, the North Sea, the Baltic and the Mediterranean. In the Black Sea, it occurs only in the vicinity of the Bosporus. The Danube catchment area is being artificially stocked with young eels brought from the estuaries of large rivers along the European Atlantic coast usually from France. As the Black Sea is far from the Atlantic, it can be assumed that the eels reach the Mediterranean and the Black Sea and then enter the Danube only in years when the

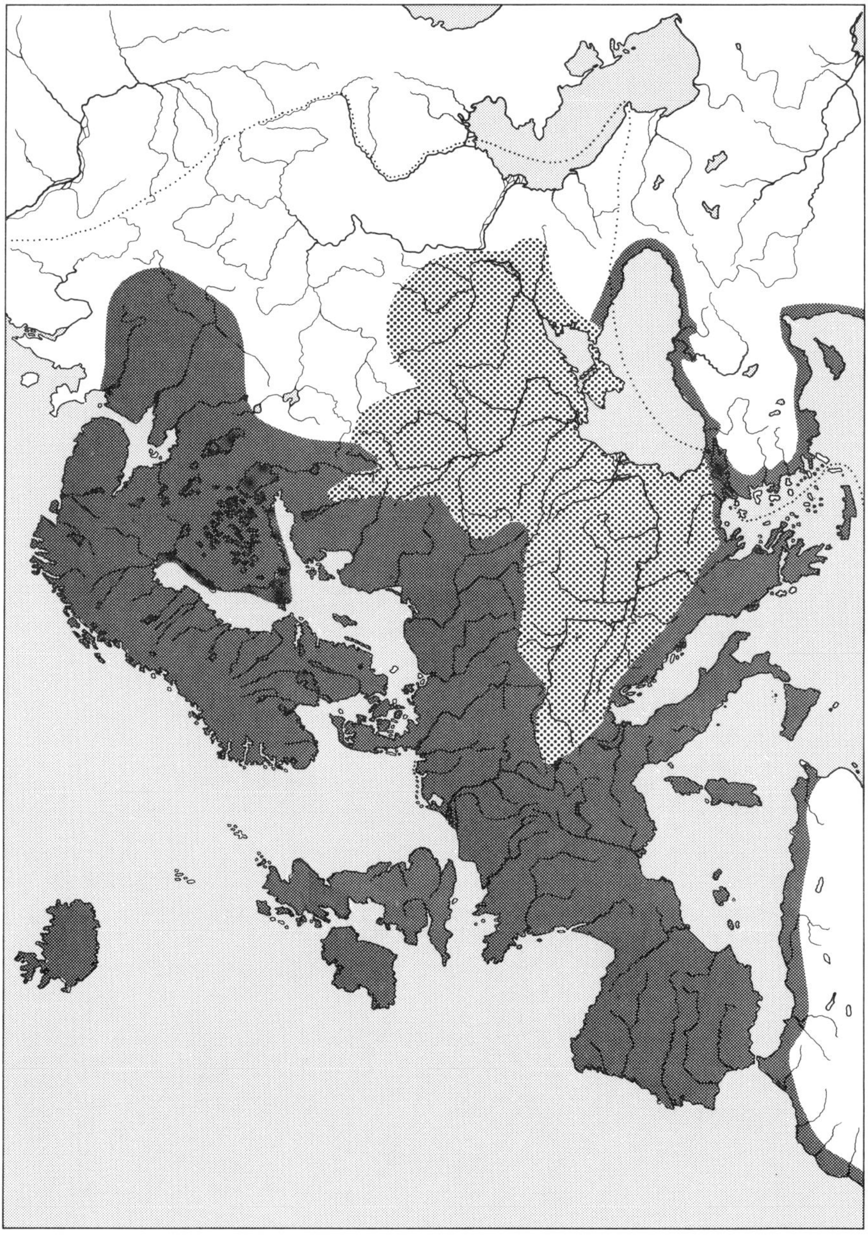

Fig. 8: Distribution of *Anguilla anguilla* in Europe. Dark area: Natural distribution area. Dotted area: enlarged distribution by stocking.

run of elvers is extremely strong. Many important observations were made before stocking was intensified. For more details see Berg (1933) and Tesch (1972). The distribution area is shown on page 64, by dots is indicated the distribution by means of stocking.

Ecology

After spawning in the Sargasso Sea, the larvae drift passively towards the European and North African coasts. A decisive role is played by the warm Gulf Stream. After three years of drifting toward Europe, the eels reach the estuaries. The sudden contact with freshwater triggers metamorphosis of the larvae into an eel-like stage (elvers). They are still transparent, but pigmentation gradually increases as they move upstream against the current. The females usually develop in fresh water, while the males spend their lives near estuaries and along the coasts. The catadromous females spend some 8 or more years in freshwater and then move towards the sea. In the estuaries the females are joined by the males and they actively swim toward America. Migration theories, possible migration routes and the mechanism of orientation are reviewed by Tesch (1972). In freshwater, the eel is an omnivorous fish; the larger the individual, the more vertebrate food it takes.

Population in Europe

The eel depends nowadays on stocking, as its natural passage is blocked by a variety of artificial barriers. Nevertheless, the natural stocks decrease in some regions, and it is apparently already extinct from the central part of the Iberian Peninsula (Cervia, pers. comm.).

Reasons for decline

The apparent stability of the eel population does not mean that it can always be managed commercially in the same way as it is now. There is evidence that catches of young eels in the estuaries of northern European rivers (e.g. Weser, Elbe, Ems) have dropped considerably. Massive migrations through estuaries as known from the past become seldom nowadays. It is believed that overfishing of juvenile eels and pollution of river mouths have contributed to the decline. Nothing is known about influence of changes of water quality at the spawning sites and during larval drifting in the sea. At the peak of reverine pollution in the 60s and 70s the eels were locally wiped put completely, eg. the Rhine and Main Rivers.

Conservation measures taken

Administrative measures to regulate catches are under consideration.

Conservation measures proposed

It is strongly recommended that, at least in some rivers, juvenile eels should have free access from the estuaries to the upstream stretches. Technical means for allowing them to do so (eel-passes) are generally known, but either they are not used or are not kept in functional condition.

3.6. Esocidae

Esox lucius Linnaeus, 1758

E. Pike
F. Brochet
G. Hecht

Vernacular name: –

Family: Esocidae

Classification: indeterminate to vulnerable

Distinguishing characteristics

The pike is characterized by its large, broad and flattened mouth with large strong teeth. The anal and dorsal fins are close to the tail fin. The colour varies, but is usually greenish-brown, the flanks being spotted with yellow-gold; young individuals are more conspicuous coloured than old ones. Meristic data: D 19–23; A 16–21; 110–130 scales in the lateral line.

Worldwide distribution

The pike is widely distributed across the northern hemisphere. The genus *Esox* has five species: one of them, *Esox lucius,* has a circumpolar distribution. On the Eurasian continent only two species occur: *E. lucius* and *E. reicherti* (Dybowski 1869).

Distribution in Europe

Commonly distributed throughout Europe with the exception of the Iberian peninsula, Iceland, Italy south of the Tiber, the streams of southern Dalmatia (Yugoslavia) and the Crimea. It is also absent from Norway north of 60°. It has been introduced into Spain and recently found in the Guadiana River of Portugal (Almaça 1965). Only the original range, without introductions is shown in the distribution map, p. 68.

Ecology

The pike is a non-migratory predator living in rivers, lakes and adjacent backwaters. It is a phytophilous spawner adapted to laying its eggs on branches, twigs, submerged plants or benthic plant debris. It is an early

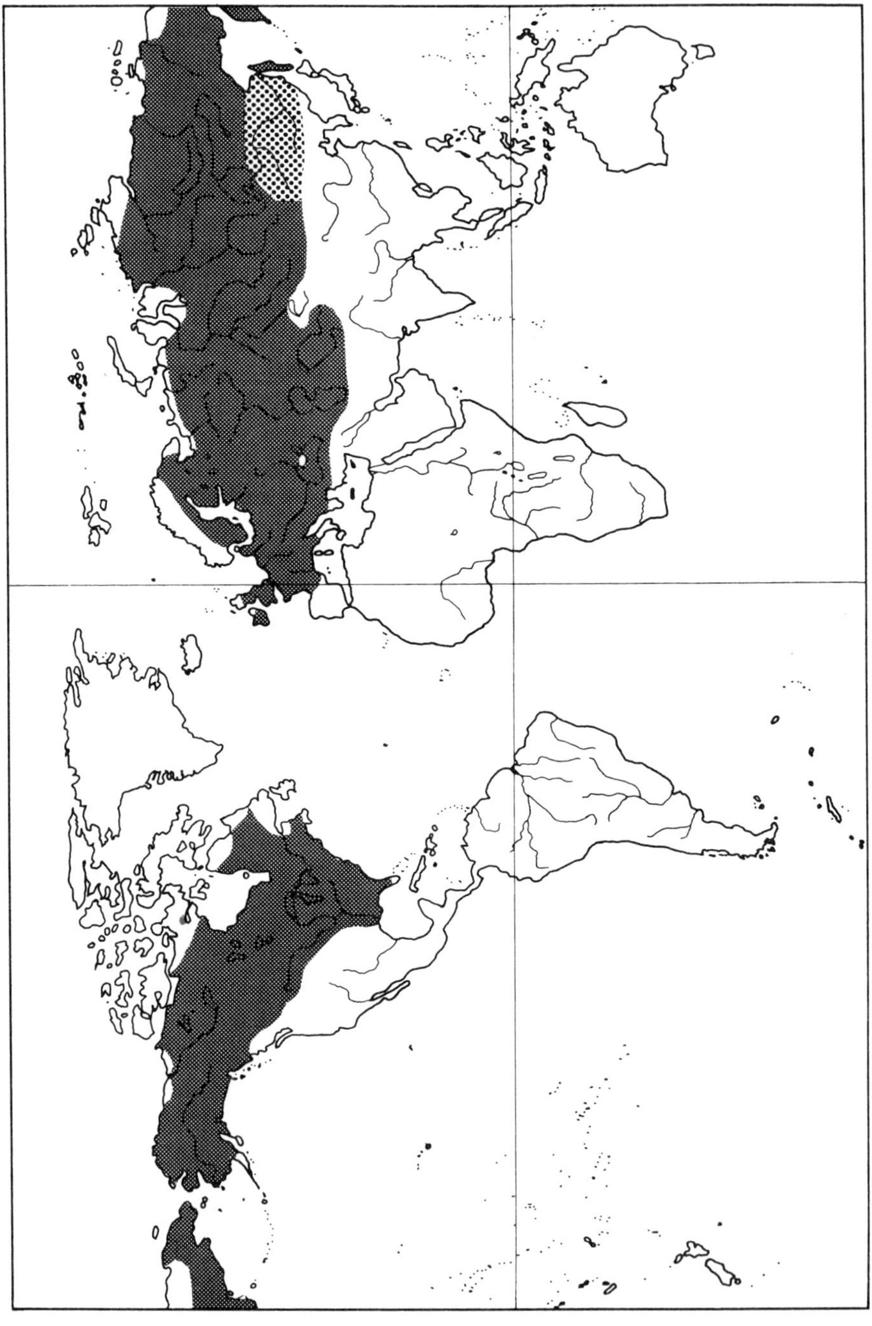

Fig. 9: Worldwide distribution of *Esox lucius* (dark area) and *E. reicherti* (dotted area).

spawner: When the substrate is suitable, it can spawn shortly after the ice melts, or sometimes later when the first vegetation emerges. Planktonic food is required only during the very earliest stages of life. When the young fish is from 4 to 7 cm long, its main food supply consists of the larvae of fishes, mostly cyprinids, that have spawned later in the season. Under suitable conditions, the pike grows rapidly and may reach well over 20 cm during its first year. Maturity is reached in the third year. It is known that pike may attain a length of over 100 cm. As the pike is one of the last predators left in European waters, it may contribute substantially to maintaining balanced multi-species fish populations.

Population in Europe

At present, pike populations are maintained mostly by stocking. As angling pressure is usually high, populations of pike are often unstable.

Reasons for decline

There are three reasons for the decline of pike, except the water quality itself, which may have some influence, follows the lack of suitable spawning substrate and heavy overfishing.

Conservation measures taken

Conservation measures usually involve artificial stocking, imposition of size limits for catch and closed seasons, but these are not sufficient. Pikes are often eleminated from closed and managed water bodies by overfishing. Restocking with undersized pike may not always be successful, particularly when this is done in early summer, because if the fish are hurt at release, they may easily die.

Conservation measures proposed

Suitable areas within a water body such as part of a lake, a bay or a pool connected with the main stream should be maintained as regeneration zones, where all fishing is forbidden. Stocking is a good substitute for natural reproduction. Deployment of artificial spanning substrates should be widely encouraged.

3.7. Umbridae

Umbra krameri Walbaum, 1792

E. European mud-minnow
F. –
G. Hundsfisch

Vernacular name: –

Family: Umbridae

Classification: vulnerable to endangered

Note: One subspecies from the Dniester has been described – *Umbra krameri pavlovi* (Kux and Libosvárský, 1957).

Distinguishing characteristics

The European mud-minnow is a small fish with a rounded snout, small jaws and relatively small teeth. It has large scales on the head and body. The dorsal fin is set far back over the insertion of the pelvic fins. The pectoral fins join the body ventrally. The body colour is dark, and sometimes an anthracite-bluish tinge on the flanks and scattered irregular brown spots on the back and sides are present. Meristic features: D 15–17 (18); A 6–9; scales in lateral line (33) 34–35 (36).

The subspecies *Umbra krameri pavlovi* shows very little meristic difference from *Umbra krameri.* Its main distinction is its longevity, which may be as much as five years.

Worldwide distribution

The genus *Umbra* is represented both in Europe and America by 3 species. In central and eastern North America are found *Umbra limi* (Kirtland, 1841) and *Umbra pygmaea* (De Kay, 1842). The latter has been introduced at several locations in Europe: France, Holland, Belgium and Germany (near Frankfurt and Hamburg). The European mud-minnow occurs only in southeastern Europe.

Distribution in Europe

Scattered distribution along the Danube from Vienna to the delta, Lake Neusiedler (Hungary and Austria), Lake Balaton (Hungary) and the lower

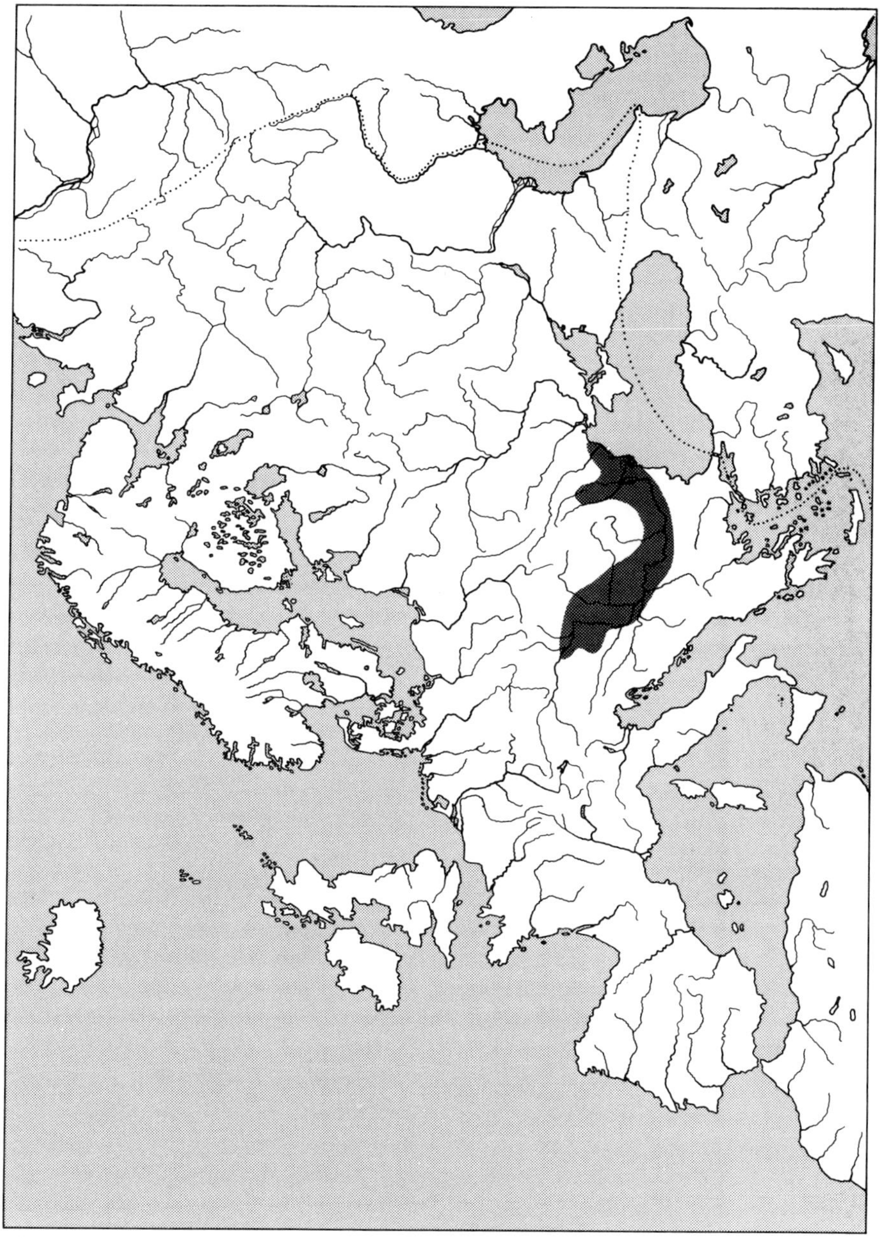

Fig. 10: Distribution of *Umbra krameri* Walbaum, 1792.

reaches of the Prut and Dniester. The last record from Lake Neusiedler dates from 1961; since then, its occurrence is very uncertain (HACKER and MEISRIEMLER, 1974). The Danube Delta and the Prut and Dniester contain the subspecies *Umbra krameri pavlovi*. See distribution map, p. 71.

Ecology

This small fish lives in pools, channels, oxbows and stagnant waters with a dense growth of plants such as *Elodea canadensis, Ceratophyllum submersum, Nymphea alba, Phragmites communis, Salvinia natans, Trapa natans* and *Stratiotes aloides.* It can withstand extremely low oxygen concentrations, as it has, in addition to gills, both alveoli and absorbing and secreting organs in the air bladder. Habitat changes, particularly reduction in the vegetation and the production of water currents by the construction of such work as irrigation channels, may be detrimental to this species. Its environmental specialization explains its scattered distribution along the Danube. This species has a short life, dying in its third year. It is territorial, feeds mostly on larger crustacean plankton and insect larvae and may be considered useful for mosquito control.

Population in Europe

The specific ecological requirements of this species has caused it to be scattered through its distribution area. Furthermore, changes in the landscape of central and eastern Europe, especially the draining of wetlands to increase agricultural production, have brought about a critical decrease in its numbers during the last 15 years. In some places, such as Slovakia and Austria, this species has almost completely disappeared.

Reasons for decline

Landscape changes and drainage of suitable sites. Chemical treatment of surrounding areas cannot be ruled out as a possible cause.

Conservation measures taken

Not known.

Conservation measures proposed

Some pools and other sites of its present and past distribution in central and eastern Europe should be preserved and maintained in their original state. Changes which might affect the hydrology and alter the vegetation must be avoided. Reintroduction is strongly recommended.

3.8. Coregonidae

General notes on the genus *Coregonus*

The taxonomy of coregonids is probably more complicated than that of any other Eurasian freshwater fish genus. The species considered here should be referred to as the "*Coregonus lavaretus* complex". This group is native to some lakes in the British Isles, to the catchment area of the Baltic Sea to the east, and to some Alpine lakes.

There is much speculation about the great diversity of the numerous closely related taxa. The most likely theory (THIENEMANN 1950) is that the primary taxon originated in Siberia, from where it spread through the whole holarctic, circumpolar area during the inter-glacial periods. As it is primarily a cold-stenotherm species, it migrated along the margins of the glaciers. This may possibly be how it arrived as far southwest as the Alpine lakes. In northern Europe, it was present at the end of the Tertiary Era. In the following interglacial period, the coregonids moved from the sea to their spawning places along the glaciers (STEINMANN 1951).

This very brief account of the historical occupation of the distribution area may provide some explanation of this taxon's great variety and to explain why it survived only in the cooler waters of northern Europe and in the Alpine lakes, often landlocked, where populations are isolated from others. For details see THIENEMANN (1921, 1922), WAGNER (1941), STEINMANN (1950, 1951) and SVÄRDSON (1957). In spite of great efforts to classify this and other related taxa, much work remains to be done as variations are extremely frequent both within and among the species. Despite extensive research, systematic taxonomists in no way agree on how these variations should be interpreted. In this study, an approach suggested by WHEELER (1969) has been adopted which assumes that each population may be somewhat different, but no attempt is made to impose a rather arbitrary system of nomenclature that cannot satisfactorily cope with such a complex situation. The general patterns of coregonid life-cycle have many similarities, so it is not so important for each particular taxon to be labelled with a sometimes dubious name, and suggestions for the future conservation of coregonids will apply to the entire complex.

The recent study of coregonids, based on electrophoretic comparison, seems to be very useful for the clarification of some genetic relationships (see FERGUSON *et al.* 1978). Its results *(Coregonus pollan – C. autumnalis)* have been accepted here. Furthermore, it may be suggested that similar studies should be undertaken of the coregonid populations in the broad alpine and subalpine area, with a view to proving or disproving the views given in the excellent work of STEINMANN (1950).

Coregonus albula Linnaeus, 1758

E. White fish, Vendace
F. Petite Marene
G. Kleine Maräne

Vernacular name: –

Family: Coregonidae

Classification: endangered

Distinguishing characteristics

C. albula resembles a herring, but zoologically is close to the salmonids. It has a relatively small head, small toothless mouth, adipose fin very close to the deeply forked tail, lateral line clearly visible. It has a pointed head, with a conical rather sharp snout, mouth terminal or superior. The dorsum is blue-green, the sides and belly silvery-white. The meristic features are: D 11–13; A 13–17; scales in lateral line 70–91; 36–52 very long gill rakers.

Distribution in Europe

Occasionally found in northern England (Castle Loch, Mill Loch, Derwent Water, Bassenthwaite Lake); Ireland (Lakes Erne, Neagh and Derg); in natural lakes east of the Elbe; lakes in Sweden, Norway and Finland; Lakes Ladoga and Onega; lakes in the upper Volga catchment area; in the Gulf of Bothnia, and native to Bavaria (Waginger and Tachen Lakes). Recently introduced further south, particularly into man-made lakes in the USSR, France (Lake Chauvet), Poland and Czechoslovakia. See distribution map, p. 76.

Ecology

A non-migratory fish of large natural lakes; only during spawning does it move into the mouths of rivers feeding the lake. The spawning season lasts from October to December, when the water temperature drops to 7° C. The eggs are laid on the sandy bottom and hatch in the very early spring or even in winter. Adult fishes live in schools. This species is a typical plankton feeder, exhibiting vertical migration in the summer months according to the stratification of food organisms. It reaches only moderate size (30 cm maximum) and lives for five years at most.

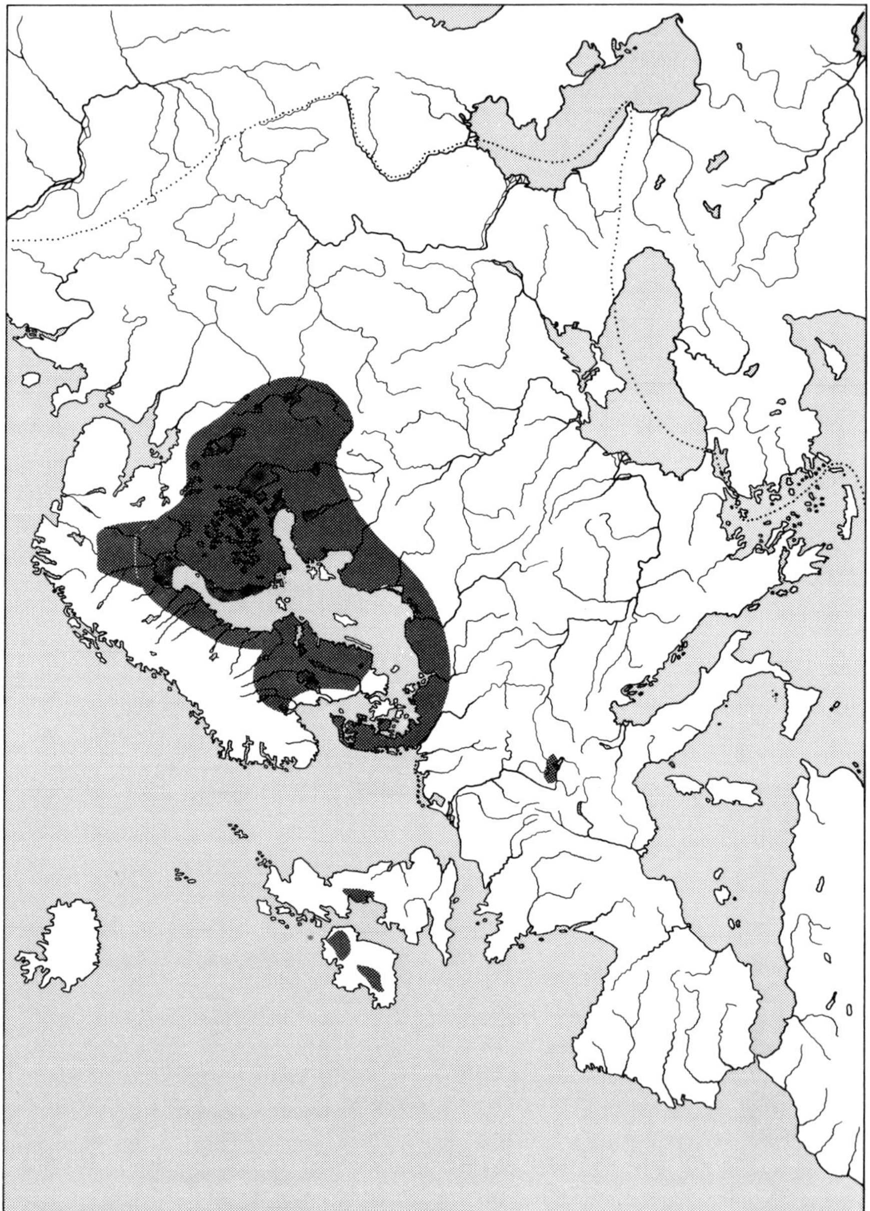

Fig. 11: Distribution of *Coregonus albula* Linnaeus, 1758.

Population in Europe

All populations in Central Europe are endangered, as the formerly oligotrophic character of the lakes is gradually changing. The water is now richer in nutrients and sometimes becomes mesotrophic. This situation may be dangerous during the winter. Moreover, increased silting often prevents the development of eggs. At present, coregonids are propagated by artificial spawning and lakes are stocked.

Reasons for decline

Changes in the trophic status of the lakes are responsible for a decline in natural reproduction.

Conservation measures taken

Although some legal measures have been taken, no practical efforts have been made to preserve the integrity of local populations.

Conservation measures proposed

Priority must be given to protecting at least some populations from management (i.e. stocking), to preserve typical autochthonous genetic material. More will be written on this subject in the general chapter on the conservation of lake populations.

Coregonus autumnalis (Pallas, 1811) (syn. *C. pollan* Thompson, 1835)

E. Irish pollan
F. –
G. –

Vernacular name: –

Family: Coregonidae

Classification: vulnerable

Distinguishing characteristics

According to its general patterns, this species is comparable to *C. albula,* but positive identification is possible only by electrophoretic methods (for details see Ferguson *et al.* 1978). It reaches a larger size than *C. albula* and does not seem to produce stunted populations.

Worldwide distribution

According to the evidence provided by the authors mentioned above, this species may be conspecific with the Alaskan *C. autumnalis.*

Distribution in Europe

The Irish pollan is confined to Lakes Neagh, Erne and Shannon. For distribution in England, see Maitland (1979).

Ecology

Similar to *C. albula.*

Reasons for decline

No decline has yet been observed, and this species is commercially exploited in Ireland (Wheeler 1969).

Conservation measures taken or proposed

No efforts for conservation are known. Fishing must be controlled. If there is additional stocking, introgression and reduction of native population may result.

Coregonus baunti Muchomedijarov, 1948

E. – Vernacular name: Varsiklöja (Swedish)

F. –

G. – Family: Coregonidae

Classification: vulnerable

Distinguishing characteristics

This taxon falls into a biological group close to *Coregonus albula,* which it resembles morphologically. The number of gill rakers ranges from (32) 36–37 (38). The primary differences are ecological.

Distribution in Europe

Occurs in three lakes in Sweden (St. Hölsjön, Asurden and Ören) and in Siberia. See distribution map, p. 80.

Ecology

This species spawns in the spring (unlike *Coregonus albula* which spawns in the autumn), at depths of 20 to 30 metres. For further details, see *Coregonus albula.*

Population in Europe

Much effort is being made to examine this species systematically and ecologically, particularly in Sweden.

Conservation measures taken: Not known.

Conservation measures proposed

The intensive research which is going on should provide a tool for active conservation measures. The release of fry foreign to the lakes inhabited by *Coregonus baunti* should be limited, preferably excluded.

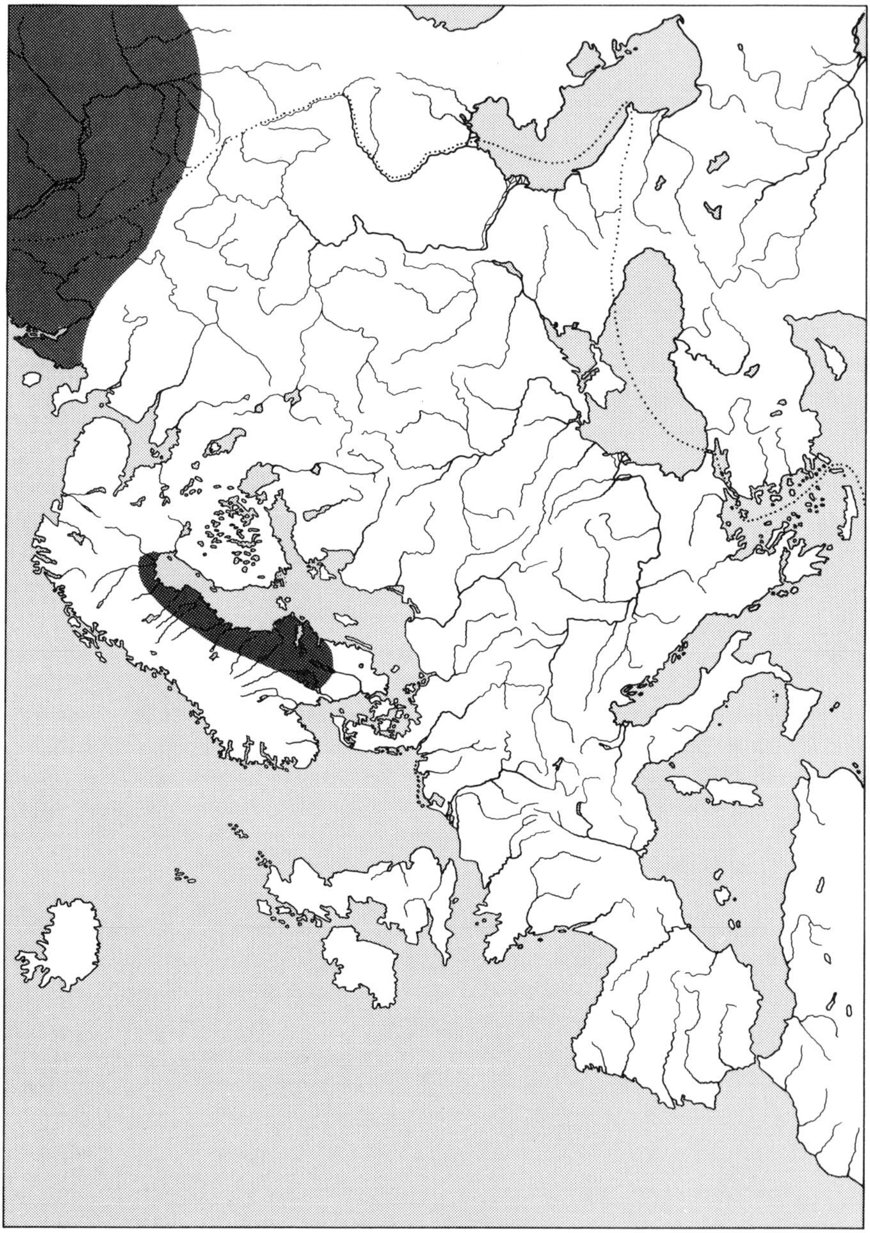

Fig. 12: European distribution of *Coregonus baunti* Muchomedijarov, 1948.

Coregonus lavaretus Linnaeus, 1758 (and subspecies)

E. Whitefish, Powan Gwymiad
F. Lavaret du Bourget, Lavoret
G. Blaufelchen, Grosse Schwebrenke

Vernacular name: –

Family: Coregonidae

Classification: endangered

Distinguishing characteristics

As a general rule, the distinguishing characteristics of *Coregonus albula* are also found in this species, but the body habitus is thicker and the position of the mouth is always either terminal or slightly ventral. Meristic data: 25–34 gill rakers, D 12–17, A 11–18, although this applies to the whole range of *Coregonus lavaretus* subspecies. For more details, see LADIGES (1965), who gives about 14 synonyms and subspecies.

Distribution in Europe

In the north of Europe and in Alpine lakes, such as Lake Constance. For distribution in the British Isles, sec MAITLAND (1979). See distribution map on page 82.

Ecology

The ecological requirements are extremely variable, depending on both the trophic status of the environment and the accompanying fish species, particularly other coregonids. Lake Constance may be taken as an example. *Coregonus wartmani* = *Coregonus lavaretus* spawns there from November to December, in deep water on sandy substrate. For details, see Nümann (1973) and Hartmann & Nümann (1977) who discribe the response of this species to environmental changes over the last decades, by review of several investigations.

Population in Europe

Common in the lakes of the Baltic Sea catchment area, as well as in those of the lower Alps. All known populations are managed, that is cropped and stocked with fry from hatcheries. The origin of the parent fishes is often un-

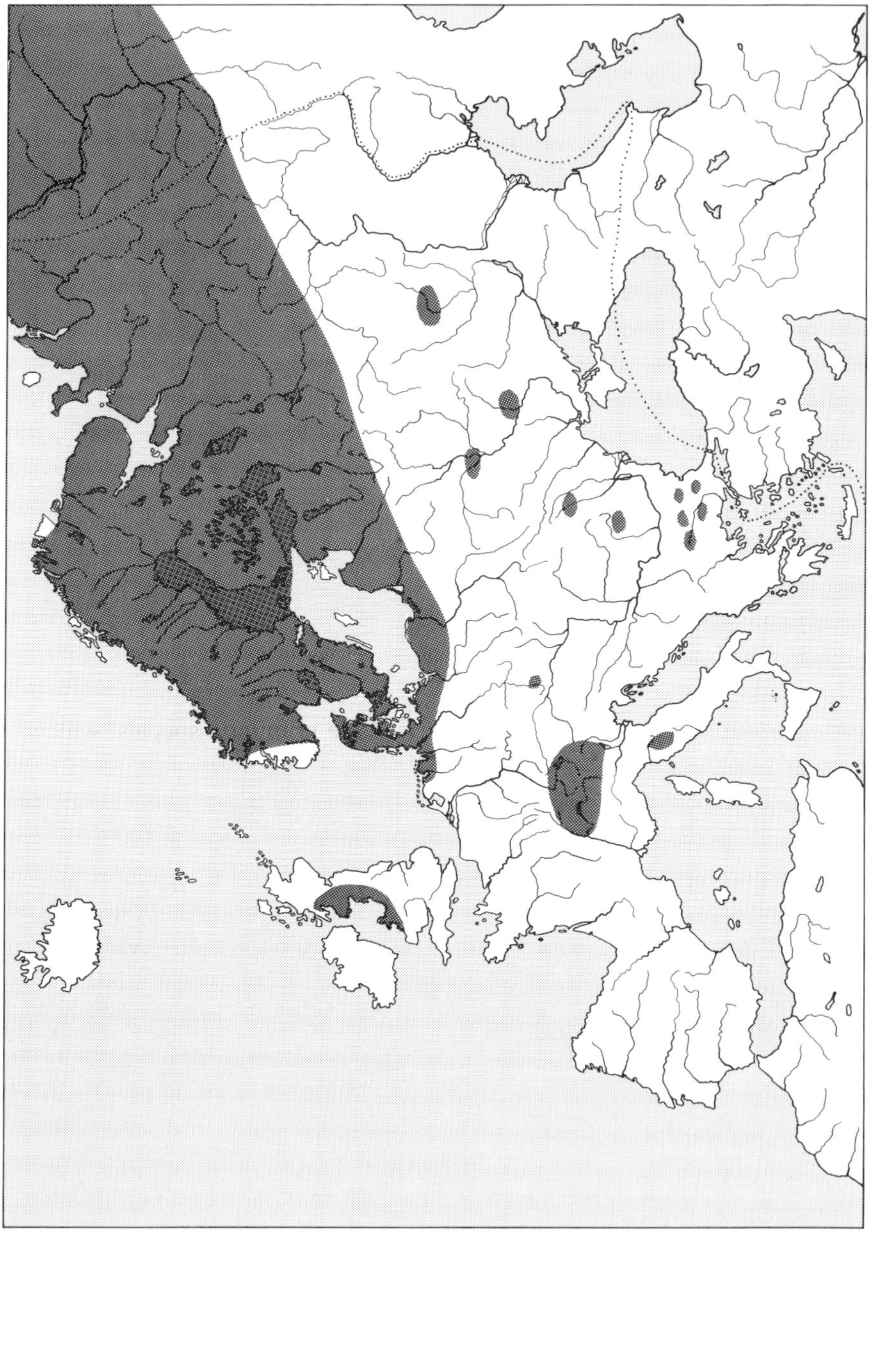

Fig. 13: Distribution of *Coregus lavaretus* Linnaeus, 1758 (including subspecies) within Europe.

known, or the fry is of mixed parentage from the whole large area. This applies particularly to the populations in the Alps. The danger of a gradual loss of genetic material will be discussed separately.

Reasons for decline

No dangerous decline has been observed. However, in lakes with increasing eutrophication, catches of coregonids in general are decreasing, though because of stocking there is no danger of extinction at present.

Conservation measures taken

At present, efforts are concentrated on maintaining the quantity of fish, so the stocking of water bodies continues. Attemps to let the fish spawn naturally at least once in its life are made by regulating the size of the gill nets mesh. Closed seasons have also been introduced. Various research programmes, particularly in Scandinavia, are focused on local taxa, in the hope of understanding their taxonomic identities as well as their life-cycle.

Conservation measures proposed

Priority should be given to the protection of the nominate species, which has been produced by environmental adaptation; stocking with fry of foreign origin should be prohibited. Continuous efforts to stock a water body, regardless of the provenance of the parent fish, may wipe out local races completely. Secondly, there must be environmental protection. Reduction of continuous eutrophication would diminish increasing silting in spawning areas.

Coregonus nasus (Pallas, 1776), [syn. *Coregonus fera* (Jurine, 1825)]

E. Broad whitefish
F. Palée du Léman
G. Grosse Maräne, Grosse Bodenrenke, Sandfelchen

Vernacular name: –

Family: Coregonidae

Classification: endangered

Distinguishing characteristics

This species bears all the patterns of coregonids, but is rather corpulent. The mouth opening is always terminal. The fish can reach a size of over 60 cm. The number of gill rakers is usually 22, but ranges between 21 and 27 in different populations.

Worldwide distribution

Occurance in cold deep water bodies in northern Europe, Siberia and Alaska.

Distribution in Europe

The distribution is similar to that of *Coregonus pidschian* (see p. 91). In Central Europe, the fish occurs in alpine lakes, in the lowlands of northern Germany, in Poland and in lakes which are connected to the Baltic Sea, e.g. Ladoga and Onega. See distribution map, p. 85.

Ecology

C. nasus lives in cold, deep lakes and is a bottom feeder. Spawning takes place either in running water or along the lake shore, often in deep water, from October to December, when the water temperature is about 2° C or 3° C, or in Siberian waters, about zero.

Population in Europe

As this species is commercially important, it is raised in hatcheries and is stocked. The taxonomic identity of particular taxa in the subalpine lakes and in Scandinavia is endangered.

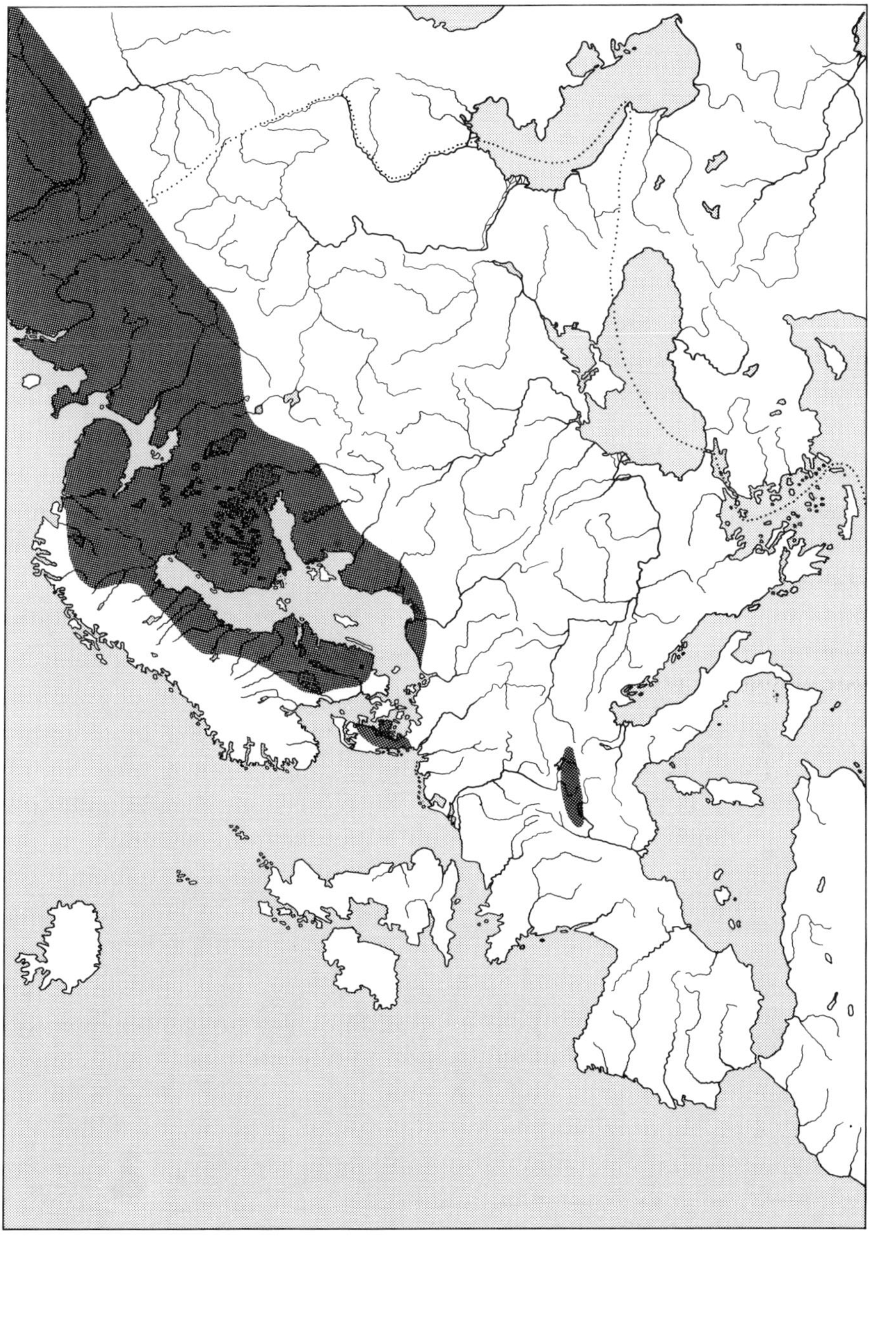

Fig. 14: European Distribution of *Coregonus nasus* (Pallas, 1776).

Reasons for decline

Decline has been observed in lakes which are losing their oligotrophic character and are silting up.

Conservation measures taken

In lakes where eutrophic conditions and silting prevent or threaten reproduction, a solution has been found in intensive stocking. As the trophic status of many lakes is very good, the fish respond with faster growth, sometimes reaching the legal size before sexual maturity. This makes stocking inevitable, if catches are to remain constant or even increase.

Conservation measures proposed

If the identity of the taxon is to be preserved, some lakes must be left without artificial stocking. Efforts to prevent organic pollution and reduction of silting rate has to be achieved.

Coregonus oxyrhynchus Linnaeus, 1758 [syn. *Coregonus macropthalmus* Nüsslin, 1882]

E. Houting
F. Bondelle
G. Schnäpel, Kleine Schwebrenke, Gangfisch

Vernacular name: –

Family: Coregonidae

Classification: endangered

Distinguishing characteristics

The houting has the usual features of this family, but also possesses a long fleshy snout, and the body habitus is more rheophilous than that of other coregonids. The number of gill rakers varies between 35 and 44.

Distribution in Europe

The general distribution stretches from the British Isles and Ireland through the marine and estuarine basins of the North and Baltic Seas, along the Russion coast and adjacent lakes eastwards to the northern Far East. Landlocked forms are found in alpine lakes, e.g. *Coregonus macropthalmus* (Nüsslin, 1882) in Lake Constance, *Coregonus generosus* (Peters, 1874) in lakes of the Oder River area and about ten forms inhabit Scandinavian lakes and Lakes Ladoga and Onega. See distribution map, p. 88.

Ecology

This species is migratory, ascending the rivers to spawn, usually in November and December. The landlocked form in Lake Constance spawns in lotic areas on sand or gravel. This fish is a bottom feeder and occasionally eats plankton. Migrations of this species in the Rhine and the Elbe ceased years ago. This migratory form is extremely rare in estuaries of English rivers, and it has never been proven that it bred there.

Population in Europe

The anadromous populations in northern Europe may be considered now as extinct, though there are records of individuals being observed as far up-

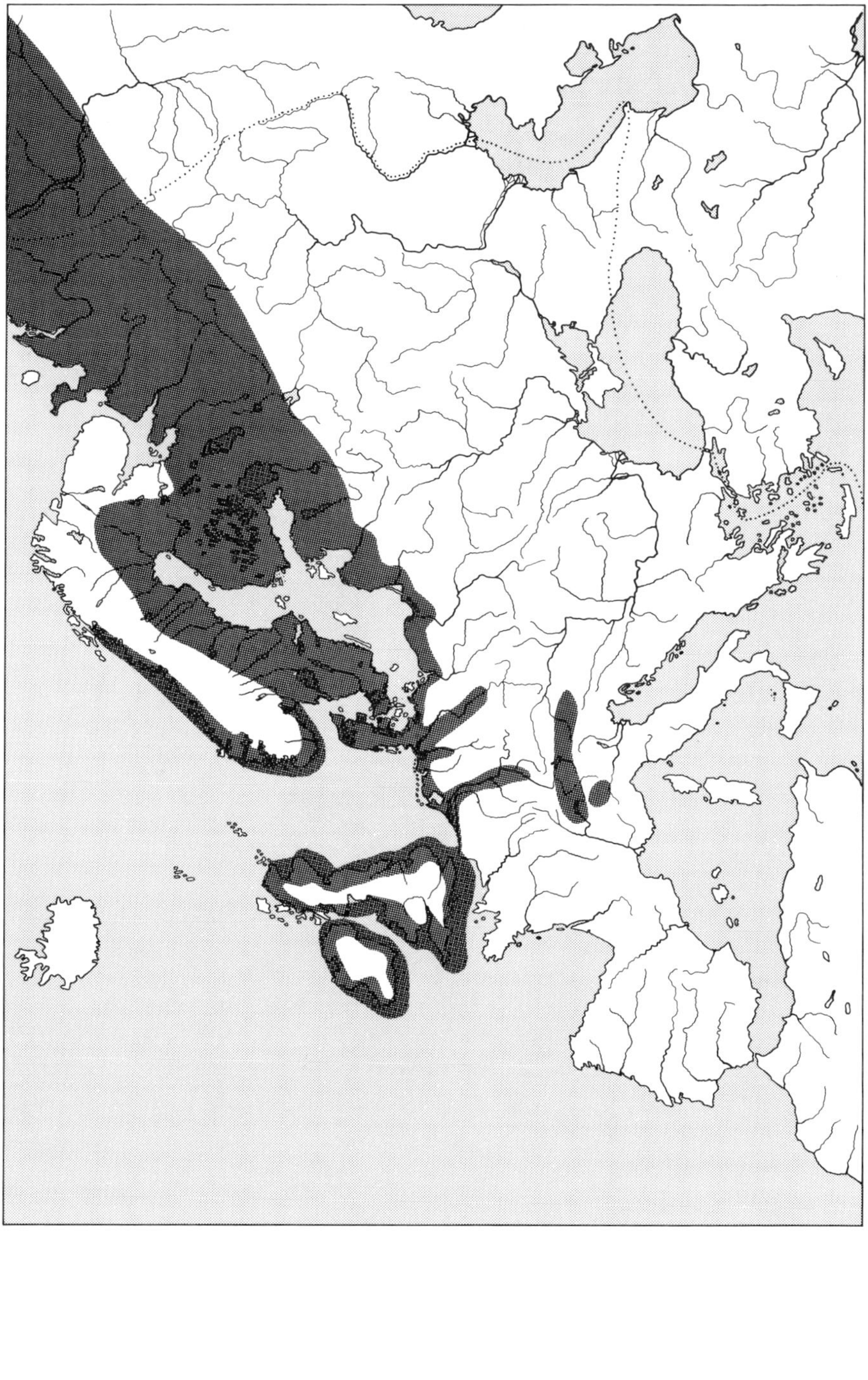

Fig. 15: Distribution of *Coregonus oxyrhinchus* Linnaeus, 1758, in Europe.

stream as Speyer, on the upper Rhine, and in the Elbe as far as northern Bohemia. Even if populations in Scandinavia and northeast Europe exist, the release of abundant fry foreign to the lakes may already have altered the identity of local taxa as in the subalpine lakes. Vagrant around the southeast coast of the British Isles. No recent records (MAITLAND 1970).

Reasons for decline

This species is seriously threatened. The reason for the decline of the anadromous forms may be the increased pollution of estuaries during the last decades, changes in water quality, works to make water courses navigable, loss of spawning grounds, overfishing and a heavy water traffic.

Conservation measures taken

See other coregonids.

Conservation measures proposed

See other coregonids.

Coregonus peled Gmelin, 1789 [syn. *Coregonus muksun* (Pallas, 1776)]

E. Northern whitefish

F. –

G. Peledmaräne

Vernacular name: Peljad (Russian)

Family: Coregonidae

Classification: endangered

Distinguishing characteristics

This coregonid with a terminal mouth belongs to the species with large body size. The number of gill rakers ranges from 44 to 68. It may reach sizeable proportions in the East (over 10 kg), but in Europe it is much smaller.

Distribution in Europe

This species occurs only in the Palaearctic Region, from the Kolyma River in the East, westwards to the Kara River. It also lives in the area south of the White Sea, in lakes in Finland, Sweden and Lapland, except the northernmost water bodies. The western range of its distribution ends at Karelia and Jutland, if populations from this region are included. During the last decades, it has been introduced into impounded waters of the Volga River.

Ecology

C. peled may be sedentary or migratory. Stunted populations from Sweden are known. It is a shore-line spawner. It may spawn in deep water in the autumn, and in other, shallow places in winter.

Reasons for decline

Eutrophication and increased acidity of waterbodies are probably the main reasons for the decline of some European populations.

Conservation measures taken

See other coregonids.

Conservation measures proposed

As for other coregonids, with particular attention to avoiding stocking with non-native fry. See *Coregonus lavaretus.*

Coregonus pidschian Gmelin, 1789 [syn. *Coregonus acronius* Rapp, 1854]

E. Humpbacked whitefish
F. Kilch
G. Kleine Bodenrenke, Kropffelchen

Vernacular name: –
Family: Coregonidae

Classification: endangered

Distinguishing characteristics

This coregonid belongs to the species with large body size. The mouth is located slightly ventrally or is almost terminal. The most important feature is the number of gill rakers which range from 15 to 25. Variation between local populations is great, particularly among populations from northern Europe and the subalpine lakes, sometimes labelled as subspecies. For the purpose of identifying individuals, consultation with ichthyologists is essential.

Worldwide distribution

This species very probably has a circumpolar distribution. However, its taxonomy is not absolutely clear. A review of taxonomical difficulties is given by Scott and Crossman (1973). It may be closely related to the North American *Coregonus clupeaformis.*

Distribution in Europe

Somewhat similar to that of *Coregonus nasus,* but it does not inhabit the rivers and lakes along the eastern shore of the Baltic. It occurs in Finnish lakes which drain into the White Sea and in arctic areas of Alaska and Siberia. It is not present in extreme northwestern Germany.

Ecology

This taxon is a cold (stenotherm) inhabitant of arctic and alpine lakes. Spawning takes place slightly later than that of the closely related species *Coregonus nasus* and may last from November to January. It is more a bottom than a plankton feeder. Its ecology is generally similar to that of *Coregonus nasus*.

Reasons for decline

See other coregonids.

Conservation measures taken

See other coregonids.

Conservation measures proposed

See *Coregonus nasus*.

3.9. Thymallidae

Thymallus thymallus (Linnaeus, 1758)

E. Grayling
F. Ombre commun
G. Äsche

Vernacular name: –

Family: Thymallidae

Classification: vulnerable

Note: The validity of the subspecies of the genus *Thymallus* Cuvier, 1829 has not been adequately demonstrated. For systematic information and literature see Scott & Crossman (1973) and Pivnička & Hensel (1978).

Distinguishing characteristics

This species can be distinguished immediately by its large dorsal fin. The dorsal rays are long; the base of the fin is larger than the length of the head. It has an adipose fin. The scales are large, set in rows, with a violet sheen. Meristic data: D 17–24, A 10–15, scales in lateral line 74 to 95.

Worldwide distribution

The genus *Thymallus* has a circumpolar distribution. Only the grayling and *Thymallus arcticus baicalensis* (Dybowski 1874), the so-called Baikal grayling, which is pelagic and spawns in rivers feeding the lake, are important to this study. The Baikal grayling was introduced into some man-made lakes in Slovakia in 1959.

Distribution in Europe

The grayling inhabits the British Isles, with the exception of northernmost Scotland and Ireland. In Scandinavia it is found only in the South. Further east, in rivers flowing into the White Sea, rivers in the Ural Mountains and the upper reaches of the Volga tributaries. In southwestern Europe, it lives in the upper to middle reaches and some tributaries of the River Loire, Rhône and Rhine. It occurs in Switzerland, northern Italy and in tributaries of the Danube. In the lowlands, it is found in small, relatively

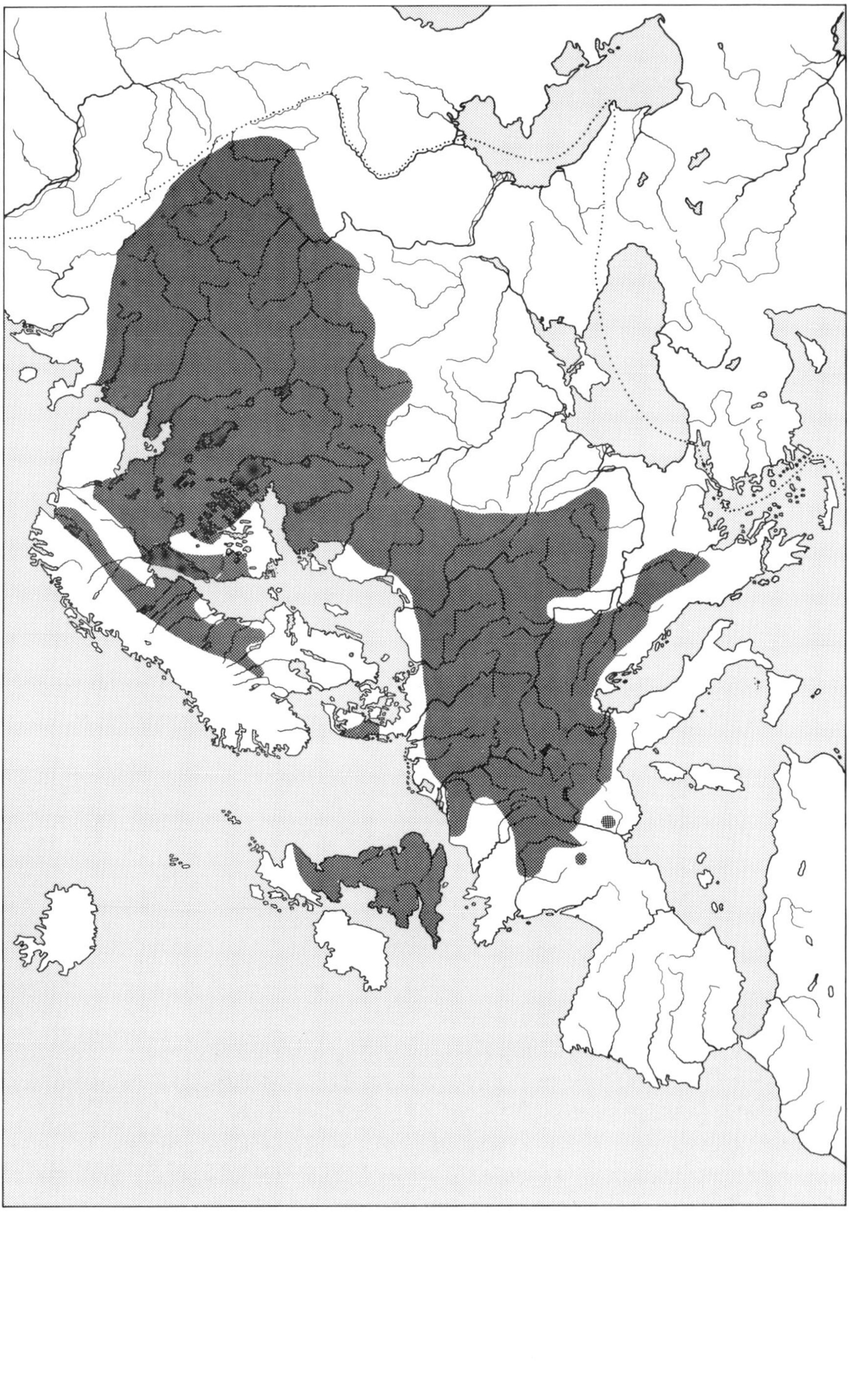

Fig. 16: European Distribution of *Thymallus thymallus* (Linnaeus, 1758).

cool streams in the Lüneburger Heide in northern Germany. Within this rather uneven distribution area, and before the pollution of central European water courses became critical, the grayling was the dominant species of the "grayling zone".

Ecology

This is a non-migratory fish of swiftly flowing streams, with deep currents over bottoms of gravel or rock covered with sand, in stretches where there is little silting. Water temperatures must not exceed 20° C for long periods in summer, and the water must always be well oxygenated. It is a lithophilous spawner. The male digs the nest, and the female, after laying a relatively small number of eggs, covers them with sand. The grayling is a drift as well as a bottom feeder. In Central Europe, it has a rather short life-span, about 5 to 6 years, but lives longer in Scandinavia. Local lake populations in the North of Europe may include individuals of over ten years old. Males become mature in the third, females in the fourth year of life. A comprehensive summary of life-history data is provided by Jankovic (1962).

Population in Europe

The grayling population is rather disrupted at present, as poor water quality and man-made changes in rivers prevent natural reproduction. It survives because of intensive stocking. Populations of the Upper Volga River disappeared after the construction of man-made lakes.

The recent re-appearance of graylings in some streams in Germany may be due to population fluctuation, which has not been explained yet and, in some places, due to improvement of water quality.

Reasons for decline

Environmental changes and poor water quality.

Conservation measures taken

Legal measures to control fishing have been taken almost everywhere, and include closed seasons and size limits. Stocking is frequent.

Conservation measures proposed

It seems fairly easy to maintain local stocks of grayling, as the species is non-migratory. Proper control of short river stretches with good spawning grounds, where there is no chance of accidential pollution, would guarantee its survival. For further details, see Chapter 4 on conservation in general.

3.10. Salmonidae

Salmo salar Linnaeus, 1758

E. Atlantic salmon
F. Saumon atlantique
G. Lachs

Vernacular name: –

Family: Salmonidae

Classification: endangered; locally threatened or extinct

Distinguishing characteristics

General characteristics of salmonids: all possess a distinct, rayless, fleshy, adipose fin on the back between the dorsal fin and the tail; the pelvic fins are abdominal and the teeth are well developed in the jaws.

In the salmon, the extreme end of the upper jaw just reaches the level of the rear edge of the eye. There are from 10 to 13 (usually 11) rows of scales between the rear edge of the adipose fin and the lateral line. There is a staggered single row of teeth on the vomerine bone, but no patch of teeth on the head of the vomer. Meristic data: D 12–14, A 9–10, lateral line 120–130, vertebrae 57–60.

Worldwide distribution

The Atlantic salmon is native to the North Atlantic region, from the Arctic Circle to Portugal in the East and from Iceland to Greenland in the West on the American continent, from the Ungava region of northern Quebec south to the Connecticut river.

Distribution in Europe

In Europe from the Mino River (Portugal) to the Pechora (USSR) in the Northeast; common in the Baltic Sea.

Ecology

The salmon is an anadromous migratory fish. It spawns in late autumn and winter, usually in November or December, though it can spawn at any time

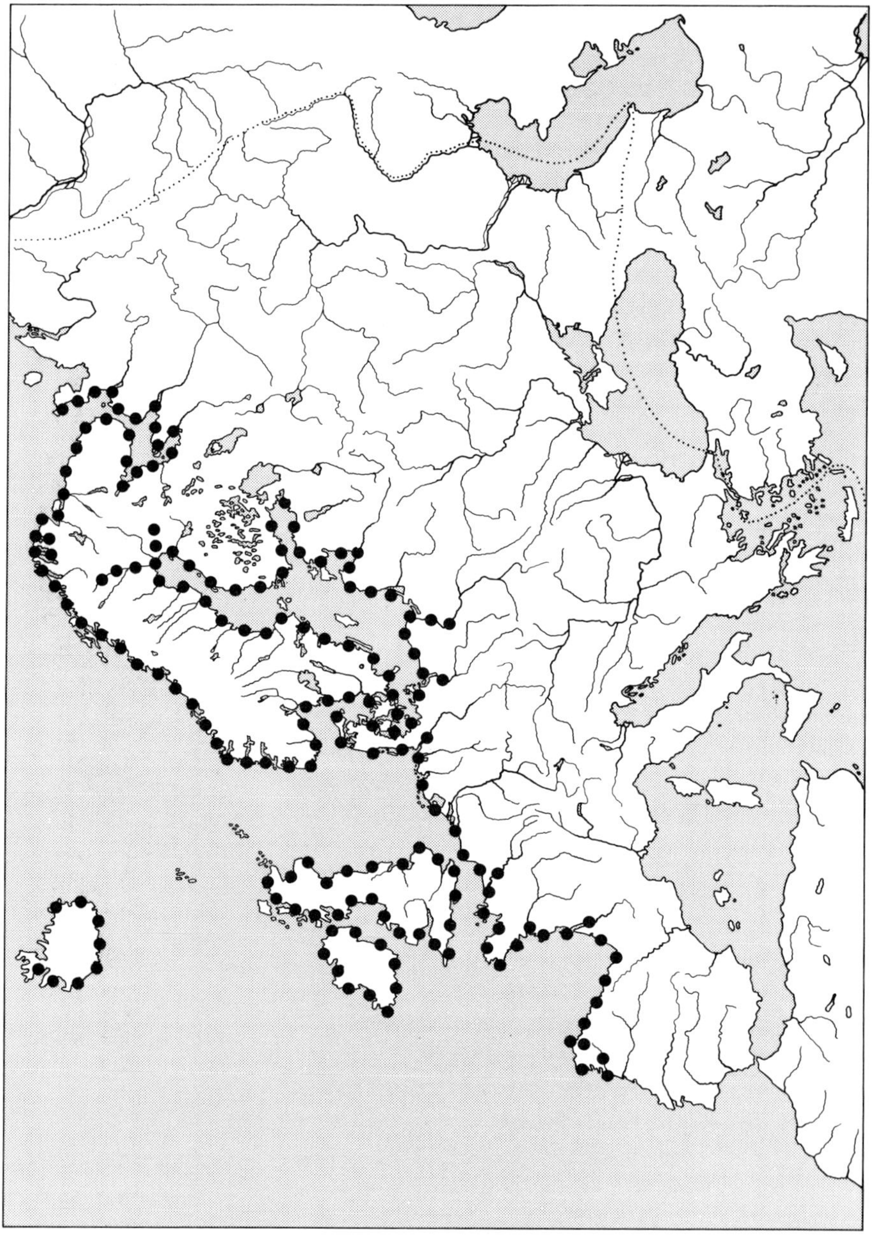

Fig. 17: Distribution of *Salmo salar* Linnaeus, 1758 in Europe.

from October till January. The female digs a nest in the gravel of the river bed. When it is big enough (depending on the size of the female) the male joins her in the nest. Immediately after the eggs are laid and fertilized, the male returns to the resting pool downstream, while the female begins to construct another nest above the first one, which has the effect of deeply burying the eggs. The spent females return to the sea almost immediately, but the males stay in the resting pools for further spawnings. Later they return to the sea and after several months (5 to 18) each fish has grown fat and ready for another spawning. On the average, about 5% spawn a second time, but only very few spawn three times (WHEELER 1969). The movements of adults are considerable. Specimens tagged in British, Swedish and Canadian waters were recaptured off the west coast of Greenland, so transatlantic journeys cannot be ruled out.

Adult salmon do not feed in fresh water. The diet of the young salmon varies greatly and is mainly a cross-section of the food available in the river. In general, growth is rapid from mid-April to the end of July. At the end of their first year of life (i.e. after one winter) only the most rapid-growing parr (juveniles) with an average length of 10 to 15 cm, can start the seaward journey. Two-year smolts (seaward moving juveniles) average 13–16 cm, while 3 year old fish may be 13 to 20 cm long.

The size of the salmon depends less on its age than on the length of time it has fed at sea. The largest fishes are those which have grown for three or four years without the physiological demands caused by spawning. They may reach an average weight of 11 kg.

Population in Europe

Stable populations are found in the British Isles (Ireland), in the countries north of the Baltic, in Norway and in Iceland. Spawning populations which used to run from the Baltic and North Seas southwards into the rivers are endangered. The salmon has completely disappeared from formerly rich rivers such as the Rhine and the Elbe. The stock entering rivers in Poland is now negligible. Recent reports about reappearance of salmon refer to the sea trouts which are often "identified" as salmons.

Reasons for decline

The main obstacles to the spawning salmon are the large construction works which have been undertaken along rivers and which prevent the fish from reaching their former spawning grounds far upstream. Pollution, particularly its continual increase since 1946, is responsible for the extinction of Central European populations. Overfishing has also played some part.

Conservation measures taken

Protection and conservation is important in those countries where salmon regularly spawns. The Scandinavian countries, the United Kingdom and Ireland protect the spawning places. Legal size limits and closed seasons have also been adopted.

Conservation measures proposed

It is hoped that protection of the environment, improved access to spawning places and pollution control in countries with good stocks of salmon can save the existing populations. Legal measures applying to fishermen might be tightened. In general, the legal limit on the size of fish might be raised and the open season shortened. Some authorities in Sweden insist that replacing environmental protection by the production of young fish in hatcheries is not the best solution. Re-stocking of some sections of the Rhine, as advocated by certain Swiss fishery authorities, may not necessarily establish permanent populations.

Salmo trutta trutta Linnaeus, 1758

E. Sea trout
F. Truite de mer
G. Meerforelle

Vernacular name: –

Family: Salmonidae

Classification: endangered; locally threatened or extinct

Distinguishing characteristics

This species has all the general distinguishing characteristics of the trouts, although it is one of the most variable species. Scales are small; there are 13 to 16 from the adipose fin to the lateral line, in an oblique series. The anal fin has seven to nine branched rays. The upper jaw reaches well past the posterior edge of the eye. The tail ist square cut. The mutual arrangement of the opercular, pre-opercular, interopercular and subopercular bones is an identifying feature as shown in the drawings of Bauch (1955) or Gasowska (1962). Colouration has only limited distinguishing value as it changes according to the habitat and stage of the animal's life. Meristic date: D 11–15, A 9–14, scales in lateral line 120 to 130.

Worldwide distribution

This species, native to Europe and Western Asia, was introduced to the American continent in 1883, where it established itself. Moreover, it has been introduced into all the continents of the world in which the environment was found suitable. For detailed information, see McCrimmon *et al.* (1970).

Distribution in Europe

The coasts of northern Europe from Cheshskaya Bay in the East to the Douro River in Portugal. In Scotland it is more common in the eastern part of the country than in the western, and it is found in all unpolluted estuaries but is less common in Ireland and Iceland. The sea trout is also abundant in the White Sea, but rare in the Neva and absent from the Pechora River. Formerly it entered all estuaries of the Baltic and North Seas.

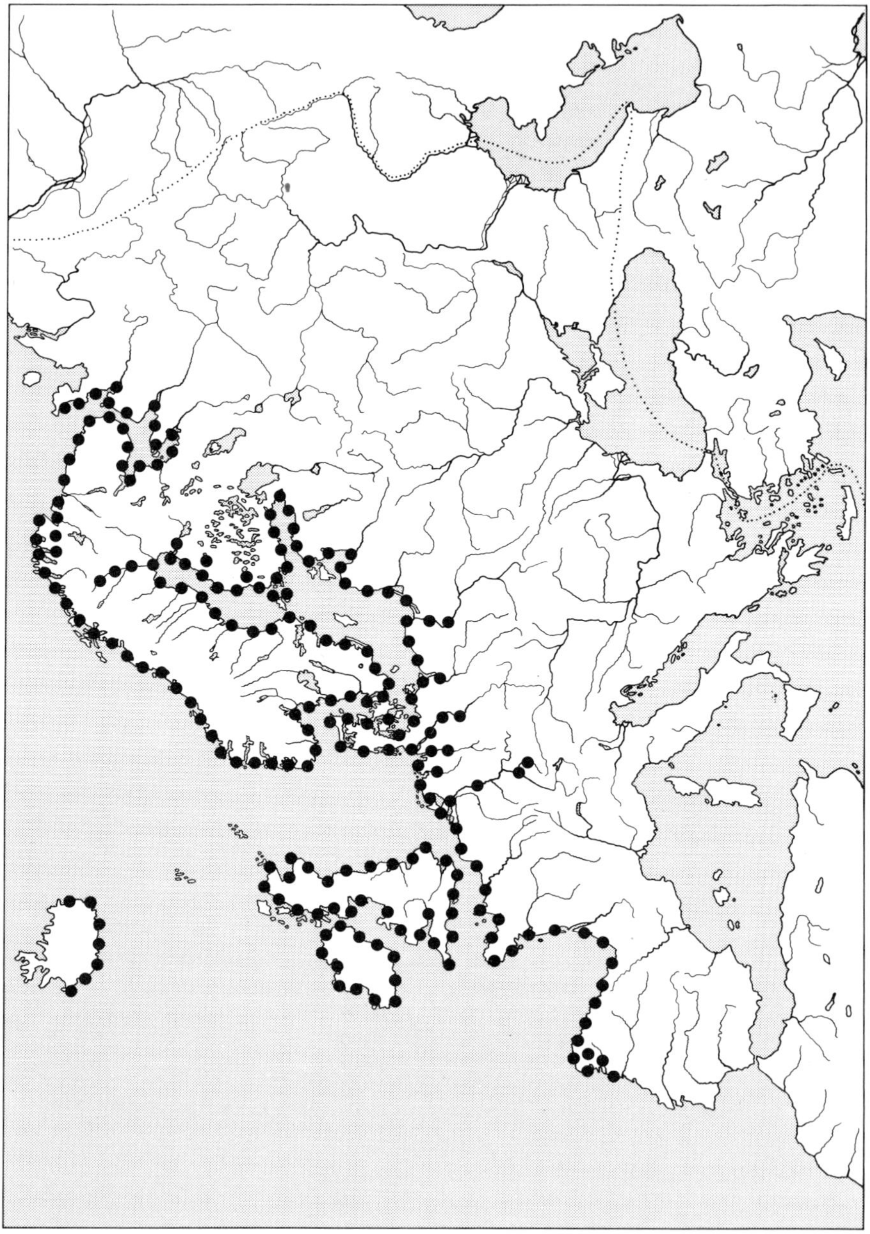

Fig. 18: European distribution of *Salmo trutta trutta* Linnaeus 1758.

Population in Europe

Salmo trutta trutta is common in the British Isles, Scandinavia and Iceland. In the rivers emptying into the North and Baltic Seas, it occurs only sporadically, because upstream migration is almost impossible. It has to be cleared whether the increasing occurrence of the sea trout in the Rhine since about 1979 is due to stocking or due to resumed natural migrations. It enters the Vistula and Oder River, but in very limited numbers. Occurance in the Rhine, Elbe and other northern European rivers are exceptions. In the Vistula River, its population is maintained artificially by means of stocking.

Ecology

The sea trout is an anadromous fish that spawns in the same places as the salmon, or farther upstream in tributaries. This species is smaller in size at maturity. The spawning run starts in summer. The spawning behaviour is similar to that of the salmon, but the neotenic parrs do not participate. The adults are less exhausted after spawning than salmon and so less likely to die, and they may spawn more often during their life span. Large-scale experiments concerning the migration of hatchery-reared and pond-kept sea trout have proved that restocking of European rivers would be feasible. For detailed information see SKROCHOWSKA (1969).

Reasons for decline: See salmon *(Salmo salar)*.

Conservation measures taken

Legal measures have been taken, and many stocking experiments were carried out. It has been proven that stocking with hatchery-reared fishes may be successful only when the passage from the sea to the point of release remains open. Released sea trouts can return to the sea provided they are conditioned by their stay as young fish in the river just before their migration to the sea and not, as was believed until recently, by their stay during the first years of life. General but useful informations including a review of literature, is well presented by SKROCHOWSKA (1969).

Conservation measures proposed: See salmon *(Salmo salar)*.

Notes on the brown trout and related taxa

Although this study mainly concerns the protection and conservation of endangered species, some attention should be paid to the nomenclature of the variety of brown trout forms. This will make communication between interested environmentalists possible. For convenience, names used are those listed in standard works, particulary LADIGES and VOGT (1965) and ILLIES (1978). Considering the 18 taxa listed, the anadromous sea trout *Salmo trutta trutta* is assumed to be the common ancestor of the great variety of present freshwater trout forms living in Europe. Their evolution was influenced greatly by the river hydrography, which was created primarily by the geomorphology and linked closely with glaciation. The distribution area of the species has split into sub-regions each usually inhabited by a land-locked form. Certain small populations survived in small isolated catchment areas, which were very little influenced by glaciation.

It is assumed that the anadromous forms of the brown trout *(S. trutta trutta)* originally occupied the Baltic and North Sea watersheds of northern Europe, as well as the Black Sea drainage area. The original anadromous form *Salmo trutta trutta,* and its landlocked form *Salmo trutta* morpha *fario*, occured in the northern distribution areas. Wherever this land-locked form established itself in lakes, the lacustrine forms designated *"lacustris"* developed. The anadromous form supposedly lived in the watersheds of the Black Sea as well. It is named *Salmo trutta labrax*. The recent occurrence of this subspecies is probably limitted only to the NE part of Turkey. A land-locked form of the taxon *"labrax"* certainly exists now, it is called *Salmo trutta labrax* morpha *fario*. The lacustrine forms, developed in the Black Sea catchment area, are similar to those in the Atlantic watersheds.

In addition to these two groups, there is a large marginal distribution area of the trout in southern Europe, which is believed to have originated from the Mediterranean "trout" which probably disappeared when the Mediterranean became warm long before the glaciation. This speculative taxon could be the ancestor of many forms living now, and collectively included under the name *Salmo trutta macrostigma* (see distribution map for *Salmo trutta*, p. 106).

At the eastern border of the "trout" distribution area, there is speculation regarding another subspecies confined to the Caspian Sea, *Salmo trutta caspius*. This one, however, may be systematically more reliable than the preceding ones, as it shows some affinity to *Salmo salar.*

For theories on the origin of the "trout", its past distribution and possible ways by which it later diverged, see BERG (1933, 1948) and the review on this topic by BALON (1969). Unless a revision of all living forms is made, together with a systematic study to find the affinity of recent taxa to their ancestors (if such a thing is possible!) all hypotheses will remain speculation, but may nevertheless be useful.

In spite of all uncertainities and inadequacies in the description of the different forms, which very often cannot be distinguished from each other, using traditional systematic methods (see *Salmo t. trutta* m. *fario* and *Salmo t. labrax* m. *fario*) it is strongly recommended that isolated forms should be protected. Recent management of waters for sport fishing may cause (or has already created) introgression, as in the case of coregonids. The most valuable habitats, mainly lakes in Yugoslavia and Albania (Ohrid and Prespa), must be protected against further stocking with non-indigenous salmonids. The same applies to the watersheds of the Baltic, North and Black Seas. Observing these precautions may help to save or to re-establish the indigenous forms that have developed in particular areas of Europe.

Salmo trutta trutta morpha *fario* Linnaeus, 1758

E. Brown trout
F. Truite de rivière
G. Bachforelle

Vernacular name: –

Family: Salmonidae

Classification: vulnerable

Note: The nomenclature used here conforms to that of LADIGES (ex. ILLIES, 1978). A critical world-wide revision of the systematics of salmonids, including trout was provided by BEHNKE (1972).

Distinguishing characteristics

There are no morphological differences between this species and the sea trout. The brown trout is only a land-locked form of the sea trout, which adapted itself to a great variety of cool, well oxygenated river environments. The colour of the adult individuals is a distinguishing factor. The brown trout is always more conspicuously coloured, usually bearing red spots that are absent from sea trout. They are also of smaller size (maximum length 40 to 50 cm). Meristic data: similar to that of the sea trout.

Worldwide distribution

This non-migratory brown trout is native to Europe, but it was introduced to almost all continents.

Distribution in Europe

The distribution of brown trout is similar to that of sea trout, stretching along the northwestern coast of Europe. Originally, the brown trout was

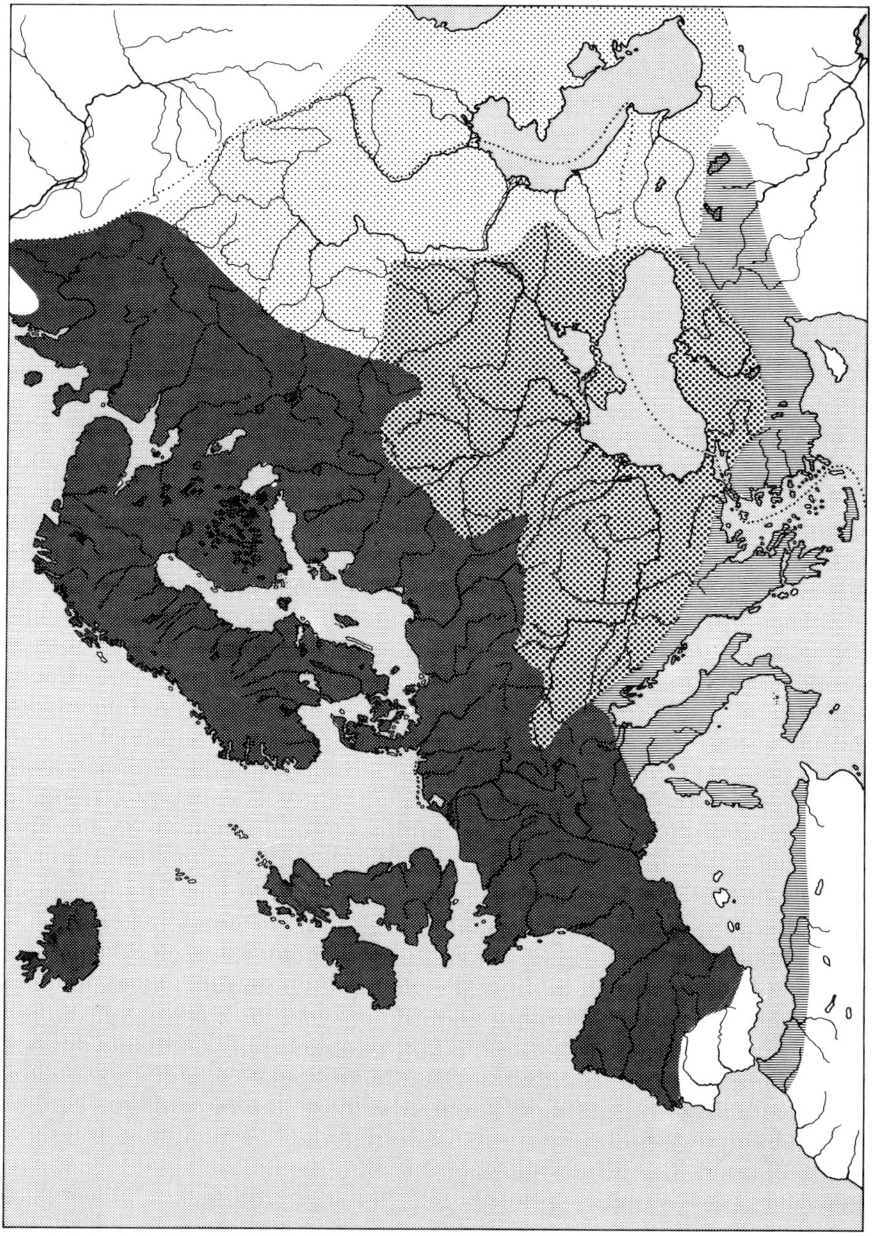

Fig. 19: General Distribution of the mostly non-migratory forms of the brown trout *(Salmo trutta)* in Europe and adjacent areas.
Salmo trutta trutta morpha *fario*, *Salmo t. labrax* morpha *fario*, *Salmo t. macrostigma*, *Salmo t. caspius.*

found in all cold, clean streams that were well saturated with oxygen in their upper and central stretches, with gradients ranging from very steep to gentle and where summer temperatures do not exceed 20° to 22° C for long periods. See distribution map.

Ecology

The brown trout is a freshwater non-migratory, cold stenotherm fish. It is found in mountain torrents and downstream to lowland rivers. Spawning begins in late autumn (in the east of its distribution area) and extends from December to January (in central and western Europe). Eggs are deposited and covered with the sand and gravel of small creeks and rivulets; hatching follows in the early spring. According to the size of individuals, brown trout can make use of all available habitats, but its behaviour is territorial. According to its feeding habits, it can be included in a category of unspecialized carnivores that consume a wide range of species: Small individuals eat the larvae of water insects while adults prey upon fish. Drifting food also makes up a considerable part of their diet. A prerequisite for the maintenance of a natural life-cycle is the availability of habitats suitable for all stages of life including spawning grounds, riffles, deeper pools, hiding places, overgrown stream banks and tiny tributaries where spawning can take place. The size of the individuals depends on food availability rather than on age. Sexual maturity is generally reached by males in the 3rd and by females in the 4th year of life. The maximum size reached is 40 to 50 cm.

Population in Europe

Brown trout occurs naturally in almost all mountain streams. In the lower reaches, which have been modified and polluted, managed populations of brown trout are usually found. In general, the more diminished the environmental, the environmental complexity by means of human alterations, the less stable the populations.

Reasons for decline

Declines in the populations of this trout are observed only locally, the reason being alterations in the environment and pollution. Most dangerous for trout populations may be artificial changes in seasonal water discharge, particularly in areas where water is taken for large-scale human consumption. During the winter, whole stretches of streams may dry out or freeze to the bottom, while during the summer months temperatures may be high. Certain management practices, particularly over-stocking and electro-fishing when it is applied too often, may also lead to the decline of populations. Furthermore, the intensive stocking with eel, practiced over the past 20 to 25 years, often in upper reaches of small streams, together with eel escapes from adjacent artificial pools, caused feeding pressure and inhibited seriously natural reproduction and recruitment of trouts. It may be considered as "eel pollution" in many places, particularly in the Fed. Rep. of Germany.

Conservation measures taken

Conservation measures are focussed on the increase in production and improvement of sport fishing. Mostly only fishable stock are maintained and little is done to conserve and to improve the environment. Fishes foreign to the watershed are used for stocking, other species are removed to ease competition, and the streams are very often overstocked. In almost all countries, closed seasons as well as bag size limits for takeable fishes have been introduced and all have proved useful.

Conservation measures proposed

To conserve the brown trout effectively on a biological basis, sanctuaries or "fish reserves" must be established, where management to increase production is prohibited. The quality of the water must also be maintained. These reserves should be located in each typical European catchment area. The Rhine, Danube and Elbe, on the Iberian Peninsula, and in selected rivers flowing to the Mediterranean Sea. This would enable the populations of different forms of brown trout to regenerate in the original habitats where they developed. This applies particularly to trouts occurring in closed catchment areas, such as the Alps and the Balkan Peninsula.

Salmo trutta labrax Pallas, 1811

E. – Vernacular name: –
F. –
G. Schwarzmeerforelle Family: Salmonidae

Classification: endangered

Distinguishing characteristics

The general habitus corresponds to that of the sea trout, but the colouration is very variable. The meristic characteristics are similar to those of the sea trout: D 13–14, A 11–12 (13), scales in lateral line 112–122. One distinctive meristic feature is the mean number of gill-rakers on the first gill arch which is usually higher (16 to 21) in the Danubian brown trout, but may overlap with counts of non-Danubian forms.

Distribution

Black and Azov Seas and the estuaries of large rivers emptying into the Black Sea from the northwest. It is supposed to be anadromous, but there is no proof of anadromous spawning migrations in the last two or three decades (BANARESCU 1964). BANARESCU observed this trout only as far as about 350 km from the Danube estuary and in the delta itself. There are no definite reports of ripening individuals being caught during ascent. It has also been observed in the lower Volga and its delta before the construction of man-made lakes. The only confirmed sites of occurence are in northeastern Turkey.

Ecology

Very little is known about the ecology of this subspecies. It supposedly ascends rivers in the spring (April to May) with gonads not fully developed and spawns in the upper reaches in the following spring. It is also believed that some individuals may spawn during the year of migration, commencing in the late autumn or winter months. Black Sea trout may reach a length of over 100 cm.

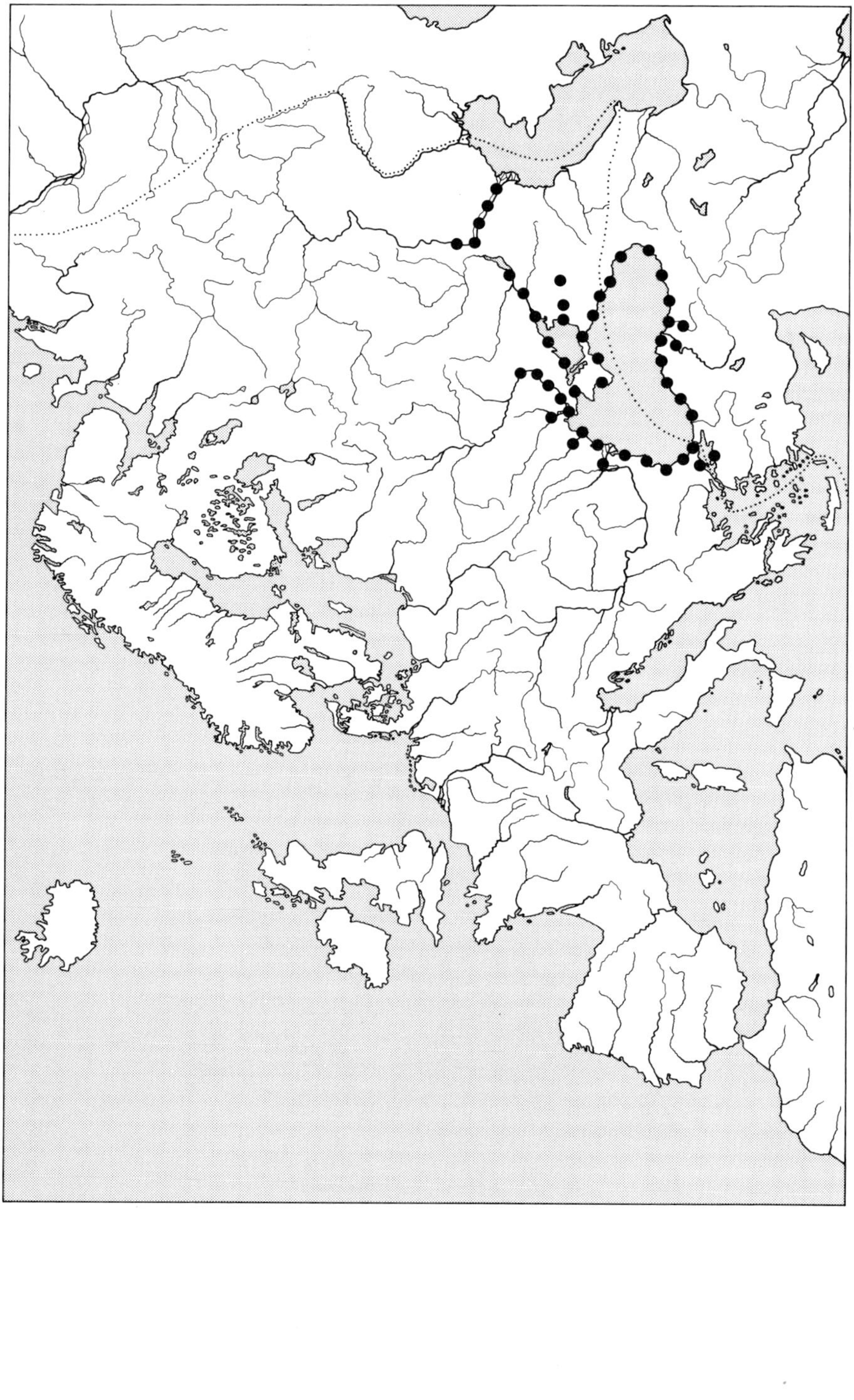

Fig. 20: *Salmo trutta labrax* Pallas, 1811: Distribution area.

Population in Europe

There are no records available about this anadromous fish. Only 7 were caught in 1939 in the Danube and it is probable that this subspecies was always rare. The population in northeastern Turkey is extremely endangered.

Reasons for decline

Not known. Adverse environmental changes and overfishing are of danger to the existing small populations.

Conservation measures taken and proposed

A list of conservation measures was provided by LELEK (1980). They include environmental protection, fishing restriction, establishing of hatcheries for this indigenous subspecies and intensive research focused on the population ecology. Reforestation efforts seem to be a prerequisit for permanent habitat improvement in areas of known occurrence in NE Turkey.

Salmo trutta labrax morpha *fario* Pallas, 1811

E. – Vernacular name: –
F. –
G. Bachforelle Family: Salmonidae

Classification: endangered

Distinguishing characteristics

For details see *Salmo trutta* m. *fario*. The only possible distinguishing characteristic is the higher number of gill rakers on the first gill arch, ranging from 12 to 20 (BALON 1967) among European and western Asian populations (LELEK 1980).

Distribution in Europe

The distribution of this race includes the Danube catchment area and rivers emptying into the Black Sea. See distribution map for *Salmo trutta*, p. 106.

For informations on populations in Europe and the ecology, see the text on *S. t. trutta* m. *fario,* as there are no differences in the life cycle of these two forms.

Conservation measures taken and proposed

See *Salmo trutta labrax.*

Salmo trutta lacustris (Linnaeus, 1758)

E. Lake trout
F. Truite de lac
G. Seeforelle

Vernacular name: –

Family: Salmonidae

Classification: endangered

Distinguishing characteristics

This subspecies of *Salmo trutta* resembles the sea trout, though it usually looks more corpulent. Spawning individuals are very often dark in colour. They may grow very large and sometimes weigh over 20 kg. There are no reliable meristic characteristics to distinguish this subspecies from other freshwater trouts.

Distribution

Its distribution is generally the same as that of the brown trout, and it is native to most large cool lakes, particularly in the Alps, Scandinavia, northern England, Ireland, as well as in the distribution area of the form "labrax" in cold lakes at higher altitudes. Modification of brown trout into the "lacustris" form has been observed in populations in some larger man-made lakes.

Ecology

Its ecology is comparable to that of the sea trout and brown trout. In late autumn, the lake trout migrates into the rivers feeding large lakes, chooses the same spawning ground as other trouts and then returns to the lakes. In some cases, spawning has been observed in the lake itself.

Population in Europe

It is difficult to assess the size of European populations as almost all are influenced by stocking with non-native fish.

Reasons for decline

No actual quantitative decline has been observed, but the integrety of the native populations is uncertain and sometimes endangered.

Conservation measures taken

In certain parts of Europe, particularly in Alpine lakes, fishes are taken from the rivers to be reared as "parent fishes" in hatcheries.

Conservation measures proposed

See the notes on the brown trout.

Salmo trutta macrostigma (Dumeril, 1858)

E. –

F. –

G. Mittelländische Bachforelle

Vernacular name: Trota macrostigma (Italian)

Family: Salmonidae

Classification: vulnerable

(See note on *Salmo trutta* morpha *fario*.)

Distinguishing characteristics

This subspecies cannot easily be distinguished from brown trout living in the northern and eastern parts of Europe. It is supposed to have fewer vertebrae than other subspecies from northern Europe, averaging 1.7 fewer. It has slightly different colouring and to some extent different body proportions. Variations according to the places of origin are apparent.

Distribution in Europe

It is found on the southern border of the trout distribution area: in the Atlas Mountains of North Africa, on Sicily and Corsica, in Greece, southern Bulgaria and eastwards to the source of the Euphrates. See distribution map for brown trout.

Ecology and conservation measures

The ecology of this subspecies of *Salmo trutta* is similar to that of brown trout. As this taxon has many local forms, particularly on the Balkan Peninsula, all indigenous populations should be protected from introgression.

Salmothymus obtusirostris (Heckel, 1851)

E. –

Vernacular name: Mekousna pastrmka (Croatian)

F. –

G. –

Family: Salmonidae

Classification: vulnerable

Distingushing characteristics

This species has all the typical patterns of salmonid fishes. The head is short and ends in a short and blunt snout. The upper jaw is wide and short, the teeth rather feebly developed. Meristic patterns: D 14, A 11 to 12, lateral line 101 to 103; 20 scales above the lateral line.

Distribution in Europe

Salmothymus obtusirostris occurs only in the streams of the Dalmatian region on the Adriatic Coast of Yugoslavia. See distribution map, p. 116.

Ecology

A non-migratory stream form, with a life-cycle similar to that of the brown trout.

Population in Europe

Specific data on populations within the distribution area are not known.

Conservation measures taken or proposed

Not known.

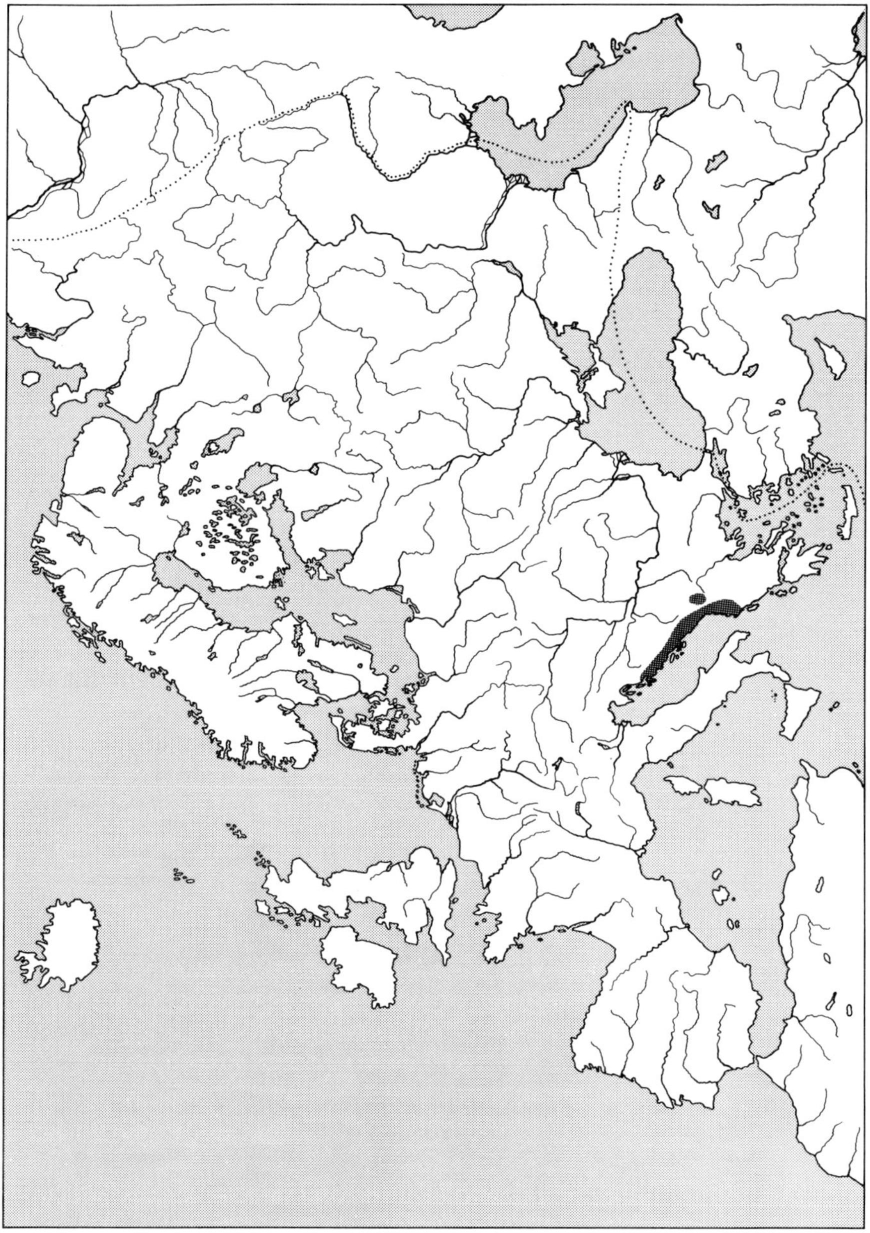

Fig. 21: Distribution area of *Salmothymus obtusirostris* (Heckel, 1851).

Salmothymus ohridanus (Steindachner, 1892)

E. – Vernacular name: Ohridska blevica (Macedonian)
F. –
G. – Family: Salmonidae

Classification: vulnerable

Distingushing characteristics

This species is distinguished from the other trout species by its silvery aspect, its hardly noticeable spots, large eyes, and the very small mouth and teeth.

Distribution in Europe and ecology

This typical indigenous fish occurs only in Lake Ohrid, where it remains usually in the sublittoral zone. It spawns in the lake and never ascends the rivers that feed it. For more details see (Stanković, 1960).

Conservation measures taken or proposed

This species is without any doubt one of the vertebrates most in need of protection in its environment. The stocking with the rainbow trout *Salmo gairdneri* into the Lake Ohrid has to be abolished.

Hucho hucho (Linnaeus, 1758)

E. Danube salmon
F. Huchon
G. Huchen

Vernacular name: –

Family: Salmonidae

Classification: endangered

Note: Only one species, *H. hucho,* with two subspecies *H. h. hucho* and *H. h. taimen* (Pallas, 1773), are important for the continent of Europe. *H. h. taimen* occurs only in the upper Volga and Pechora River systems. For details concerning systematics and taxonomy, see Hensel and Holčík (1982).

Distinguishing characteristics

This species has a typical salmonid body form, rounded in transection. The head is long, with a dorso-ventrally compressed maxillary extending well behind the eye. On the head of the vomer there are 4 to 8 hooked teeth in one row but no teeth on its shaft, and there are 16–18 rakers on the first gill arch. Meristic features: D 12–19, A 10–16, in the lateral line 107 to 194 scales, above the lateral line 25 to 38 scales. It can reach a weight well over 20 kg.

Distribution in Europe

The original range includes only the Danube catchment area. The fish was more abundant in the right-hand tributaries and was also found in the River Pruth. It very seldom inhabits the lower reaches of the streams. During the past 20 to 30 years, attempts were made to introduce this species west of its natural ranges, but they usually failed as in the case of the Thames and the Rhine. Attempts of introduction into the Elbe system also seem to have failed, as there is still no evidence of natural reproduction. The same thing has happened in the Poprad and the Dunajec tributary to the Vistula, and even in stream sections within its distribution area where it did not occur naturally. See distribution map according to Holčík *et al.* 1984, p. 119. For further details see Randik (1976), Jungwirth (1978) and Holčík (1982).

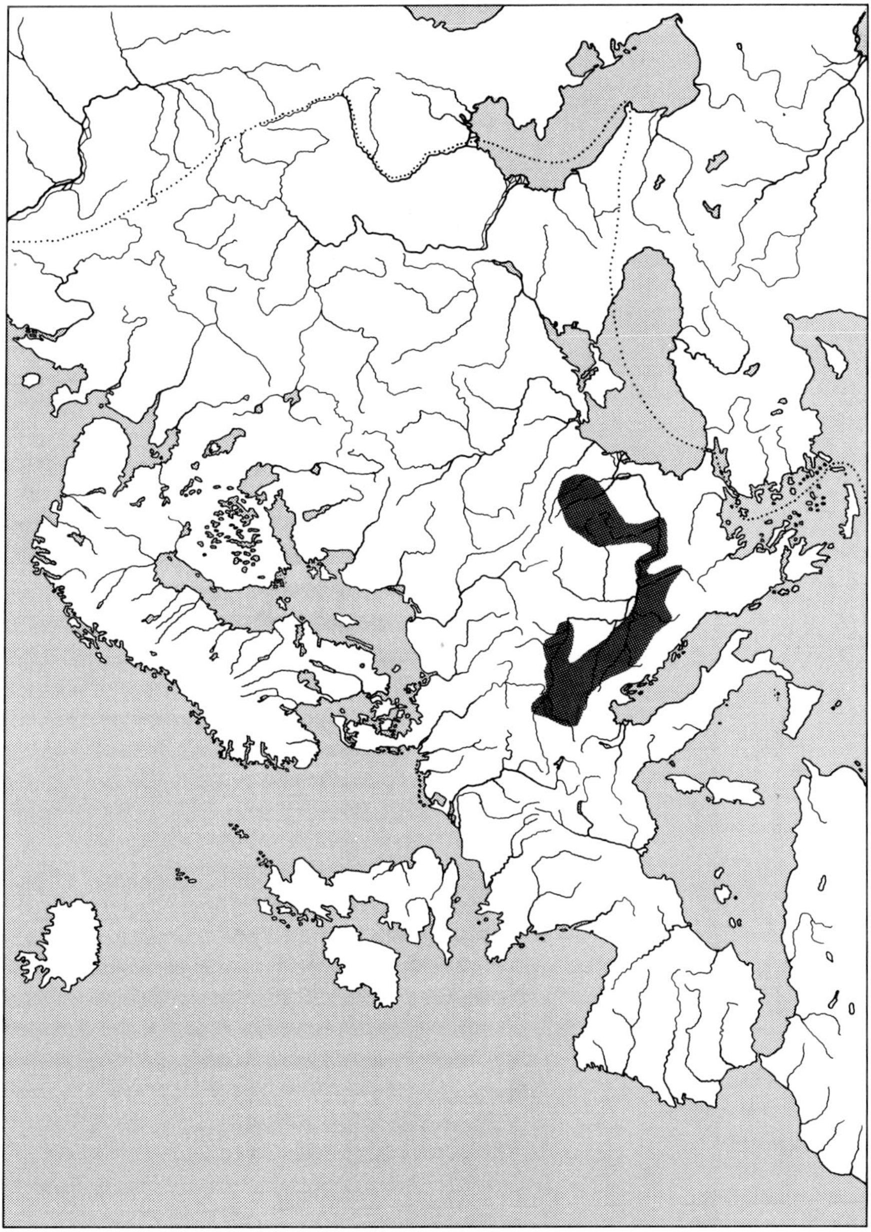

Fig. 22: Distribution of *Hucho h. hucho* (Linnaeus, 1768).

Ecology

The Danube salmon usually occupies transitional areas between the brown trout and grayling zone. Juvenile individuals live in small tributaries and, as they grow, move to larger ones. Spawning takes place on gravel bottom in mountain torrents in spring (usually mid-April). It shows some similarity to the trout in its spawning behaviour. The growth rate is very fast. By the end of the year of hatching, the fish reaches a body length of about 13 cm, and in the second year over 30 cm. At 5 years, the huchen can measure over 55 cm (KIRKA 1963). The diet of juveniles is comparable to that of trout, while adults are typical predators. If *Chondrostoma nasus* is present in its territories, it is the main food fish, as it was, before the rivers within the range of the huchen were modified. If this preferred prey is not available, it consumes other fish. A good summary of biological information is provided by PRAWOCHENSKY and KOLDER (1968), in symposium proceedings edited by RANDIK (1976) and by HARSANYI (1982).

Population in Europe

The population of *H. hucho* is seriously endangered in Europe, as there are only a few rivers in Austria, Czechoslovakia, the Federal Republic of Germany and Yugoslavia where it reproduces naturally. These habitats are often used as a source of mature individuals to supply the hatcheries. Existing populations owe their survival mainly to intensive stocking.

Reasons for decline

Since the Middle Ages, it has been the most popular salmonid species in its distribution area, and overfishing has contributed to the reduction of almost all populations. Man-made lakes in the upper reaches of rivers within its distribution area have also had an adverse effect and reduced the populations. In many cases, typical "trout stream management" of river stretches was applied and cyprinid species (including *Chondrostoma nasus*) were removed as possible competitors. This had unfortunate results, as maturing *H. hucho* depend on fish as their main food resource. Known spawning places were not always protected, and spawning pairs were often disturbed or illegally captured. Water quality changes need not be mentioned – they had the same or an even worse influence on huchen populations as on other salmonids. This species occurs not only in high altitudes, but also in lower reaches as well, except when these are too warm during the summer. Eutrophication, linked with higher seasonal and diurnal oxygen variations, probably contributed to its gradual extinction even before the massive water pollution began after the Second World War. The disappearance of this species from the March River can be seen as a case history of man-induced stress on a fish species.

Conservation measures taken

The most important conservation measure is stocking. Although this helps to maintain the species in the river, it cannot replace the natural spawning of indigenous individuals. In past and present, attempts have been made in Germany to stock rivers with fishes from Yugoslavia. German fishes have been released in Czechoslovakia before the Second World War, and Slovakian fishes have been used in Austria, Poland, and other countries. The introduction of closed seasons was very important as it is very effective for protecting the Danube salmon. In some parts of Europe, the season lasts only for a few weeks before and after Christmas. The size limit is usually set at 60 to 70 cm.

Conservation measures proposed

The best possible means of conservation is again the protection of the biotope. It is hoped that rigorous legislative measures will be applied to some remaining stretches of river, particularly in Austria, Czechoslovakia, the Federal Republic of Germany, and Yugoslavia. Existing protective measures do not ensure that this species is out of danger. Where the conservation of Danube salmon by environmental protection cannot be fully effective, hatcheries must be provided in connection to river stretches long enough to accommodate sufficient numbers of adult fishes. A supplementary practice under consideration is to keep parent fishes in ponds, as suggested by JUNGWIRTH (1978). When the introduction of the related subspecies, *Hucho h. taimen* is considered, it is strongly recommended that experiments first be conducted in an absolutely controlled environment, to prevent possible hybridization or genetic damage to the European stock. Restocking attempts within the former range of this species should be encouraged and supported financially. HARSÁNYI (1986) provided a catalogue of recommendation.

Salvelinus alpinus (Linnaeus, 1758)

E. Charr
F. Ombre-chevalier
G. Seesaibling

Vernacular name: –

Family: Salmonidae

Classification: endangered

Note: Ladiges and Vogt (1965, 1979) list 24 forms, both landlocked and migratory.

Distinguishing characteristics

The charr displays all the general features of salmonids. They have an adipose fin, pelvic fins placed directly beneath the dorsal fin and minute scales covering the body. The hind part of the mouth extends to the rear edge of the eye and occasionally, particularly in the males, beyond it. The vomer has teeth on the head, not on the shaft and there are no teeth in the centre of the palate. The colour of the dorsum is fairly uniform and the black stripe on the anal fin is absent. The colour of the freshwater forms varies greatly with locality. Meristic features: D 12–15, A 10–13, lat. line 190–240 scales.

Worldwide distribution

This species has the northernmost range of any freshwater fish. It has a circumpolar distribution and is found in inshore marine waters, lakes and rivers in arctic regions of North America, Asia, Europe, Iceland, Greenland and innumerable arctic islands. It may be anadromous or remain landlocked. Freshwater populations occur in more southerly latitudes in Europe including the British Isles, Ireland and alpine lakes. For the comprehensive literature see Balon (Ed.) (1980).

Distribution in Europe

For details see the distribution map. It is generally accepted that its distribution is linked with glaciation. Charr in the Alps are considered to be glacial relicts.

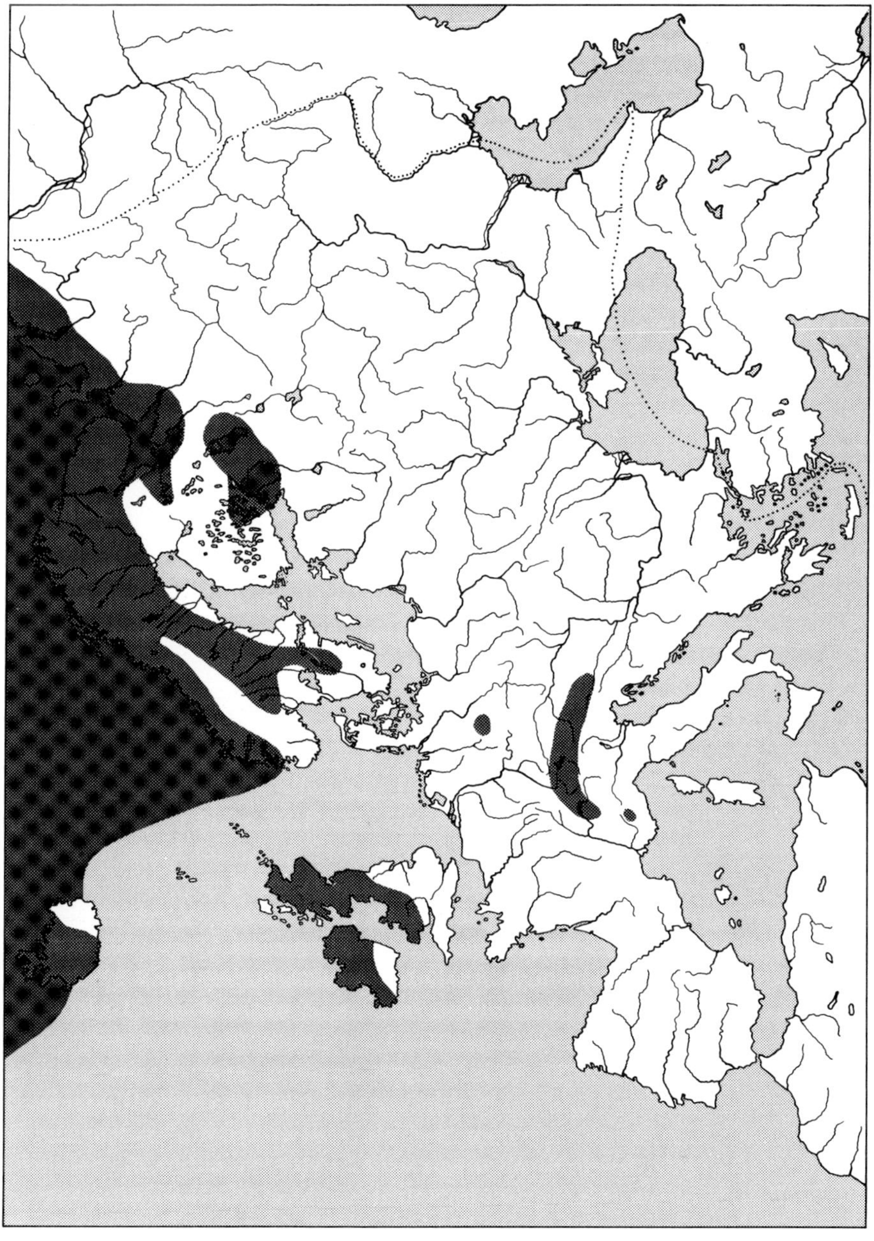

Fig. 23: European distribution of *Salvelinus alpinus* (Linnaeus, 1758).

Population in Europe

It is extremely difficult to speak of a "population" in Europe, because each water body or catchment area where the charr is indigenous is a separate unit. Basically, two different types of populations are recognized: migratory and landlocked.

Ecology

The migratory charr may grow large, but rarely lives more than 11 years. Adult individuals ascend rivers between July and September, very often after heavy rain. After spawning in early winter, they stay in the rivers until the ice breaks, returning to the sea with the first floods.

Freshwater populations occur in deep lakes in the Alps, England and Ireland. They may have different ecological adaptation to the depths at which they live and to the spawning substrate. For example, the Windermere charr in northern England has 2 different populations: autumn spawners breed during November and December in shallow water, 1 to 3 m deep, on stony ground, while spring spawners reproduce in February and March at depths from 20 to 30 m. For the charr in Bavaria and Austria, BERG (1933) accepted the distinct names listed by HAEMPEL (1930) and BURESCH (1925), of which there are at least four:

the "normal";

"Wildfangsaibling" from the largest lakes, e.g. Constance, Geneva, Mond, etc.;

"Schwarzreiter", which normally grows to a length of 14 to 15 cm;

"Tiefseesaibling", living in depths from 30 to 100 m and reaching a length of from 12 to 20 cm.

Each population has a different ecology, but it is not certain how much the ecological demands may be modified.

Reasons for decline

No actual decline has been observed and variation in catches is probably linked with eutrophication, as in Lake Constance. Locally, as is the case in Ireland, the introduction of cyprinids prooved negatively on the stock of charr.

Conservation measures taken

Protection and conservation measures, as understood in fishery management, involve stocking the lake with charr. As the introduction of large or fast growing forms are deemed desirable, small forms may disappear. Legal measures are employed, e.g. closed seasons, mesh-size limits etc. For details of the efforts to "improve" the different stocks of charr in the Alps, see NERESHEIMER (1937). The difficulties in determining the identity of many existing forms may be extremely complex because of introgression, as in the family Coregonidae.

Conservation measures proposed

It is suggested that at least some habitats at high altitude should be left undisturbed. Their mainly monospecific populations are stable for the following reasons: constant recruitment, little or no fishing pressure, no pollution. When management in the form of stocking should be started, these populations may be altered and consequently disappear because of competition with the new form. The population in Lake Dössner, east of Mallnitz (Kärnten, Austria) at an altitude of 2281 m should be permanently refrained from all stocking experiments. This is probably the southernmost locality of this species.

It is believed that closed seasons and catch restrictions may help protect migratory forms in northern Europe, as these cannot be influenced by stocking as easily as landlocked populations. Vigorous restriction should be imposed on the introduction of individuals from other regions.

Stenodus leucichthys leucichthys (Güldenstädt, 1772) [1]

E. –
F. –
G. Weisslachs

Vernacular name: Beloribitza (Russian)
Family: Salmonidae

Classification: endangered

Note: the subspecies *Stenodus l. nelma* (Pallas, 1773) is treated separately (page 128).

Distinguishing characteristics

This species is a silvery salmonid fish, resembling the coregonids to some extent in its habitus. For a description of the mouth, and other features, see *Stenodus l. nelma.* Meristic features: D (11) 14–17 (19), A (11) 14–16 (19), 90 to 120 scales in the lateral line.

Distribution

It inhabits the Caspian Sea and ascends the Volga and Kama Rivers and their tributaries. Most are found in the Ufa River. The species was also found in the Ural River and, rarely, in the Terek. The present distribution is very probably more restricted.

Ecology

Stenodus is an anadromous species. Before the construction of man-made lakes on the Volga, its migratory route totalled 3000 km. It does not take food during migration and spawns in October and November. Females mature at 6 or 7 years of age, males at from 5 to 7 years at a length of 80 to 95 cm.

Population in Europe

Not known in detail. However, the species is managed as it is of economic value. It is believed that fewer than 2,000 specimens of this fish exist.

---

[1] See Svetovidov (1968), quoted in the IUCN Red Data Book Vol. IV, 1977.

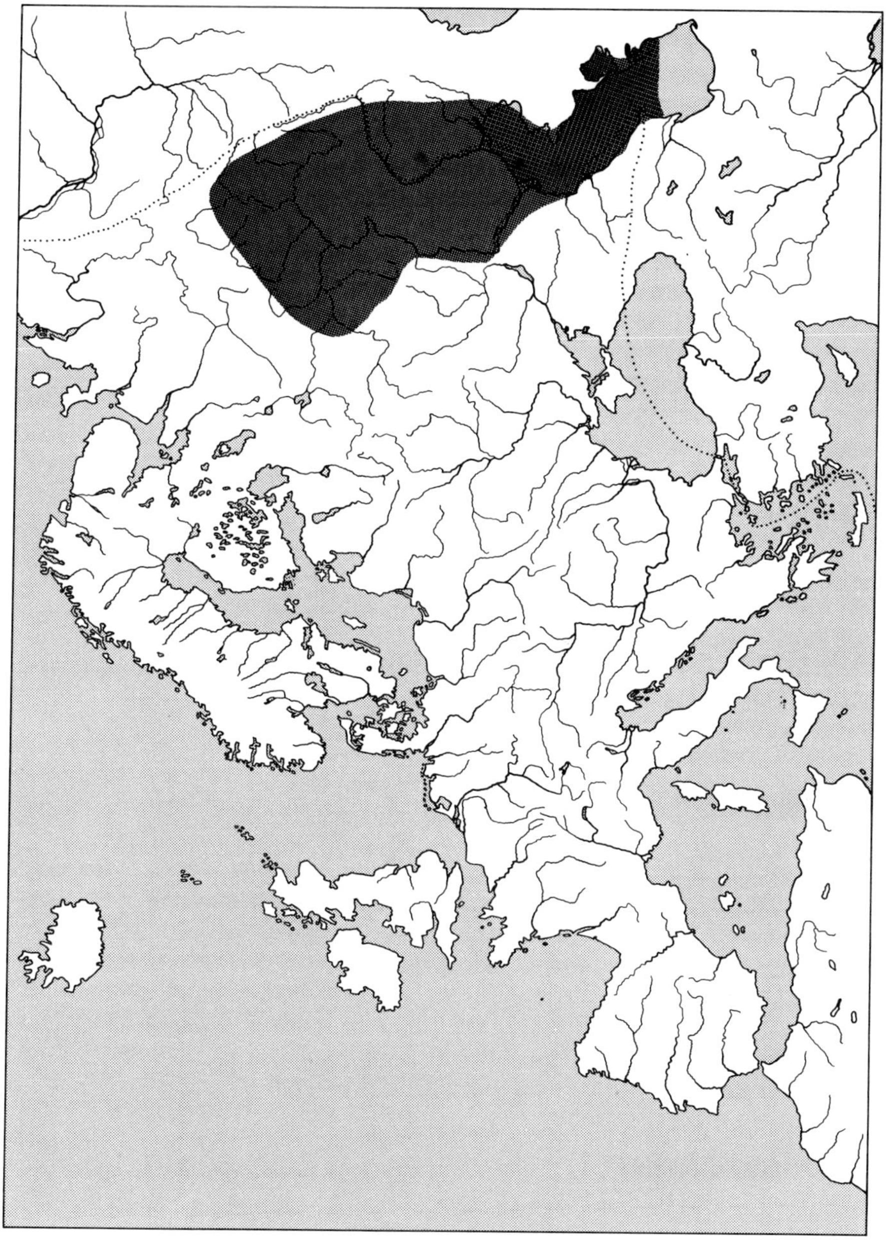

Fig. 24: Distribution area of *Stenodus leucichthys* (Güldenstädt, 1772).

Conservation measures taken or proposed

The population is maintained by stocking with hatchlings, but breeding material is supplied from the northern stock of the subspecies *Stenodus l. nelma,* which is less threatened than the Caspian species. But this well minded effort may lead to its total extinction.

Stenodus leucichthys nelma (Pallas, 1773)

E. –

F. –

G. Weisslachs

Vernacular name: Nelma (Russian)

Family: Salmonidae

Classification: intermediate

Distinguishing characteristics

Salmonid fish, silvery overall, green to pale brown on the back. Elongated body, resembles to some extent the coregonids. Mouth large, terminal, lower jaw projecting in front of upper one, fine teeth developed on anterior part of lower jaw, tongue, praemaxillaries, head of maxillaries, vomer and palatines. Meristic features: D 12 to 18, A 16 to 20, 90 to 120 scales in the lateral line.

Distribution

In the White Sea, rivers flowing into the Dvina and the Ob, and further east throughout arctic Siberia. Also in northwestern North America. See distribution map on page 129. Population in Europe not known in detail.

Ecology: Anadromous fish, spawns in rivers, lives in the sea and in rivers.

Reasons for decline: There is a danger from overfishing.

Conservation measures

Artificial propagation in hatcheries is being practised. No conservation measures linked with habitat protection are known.

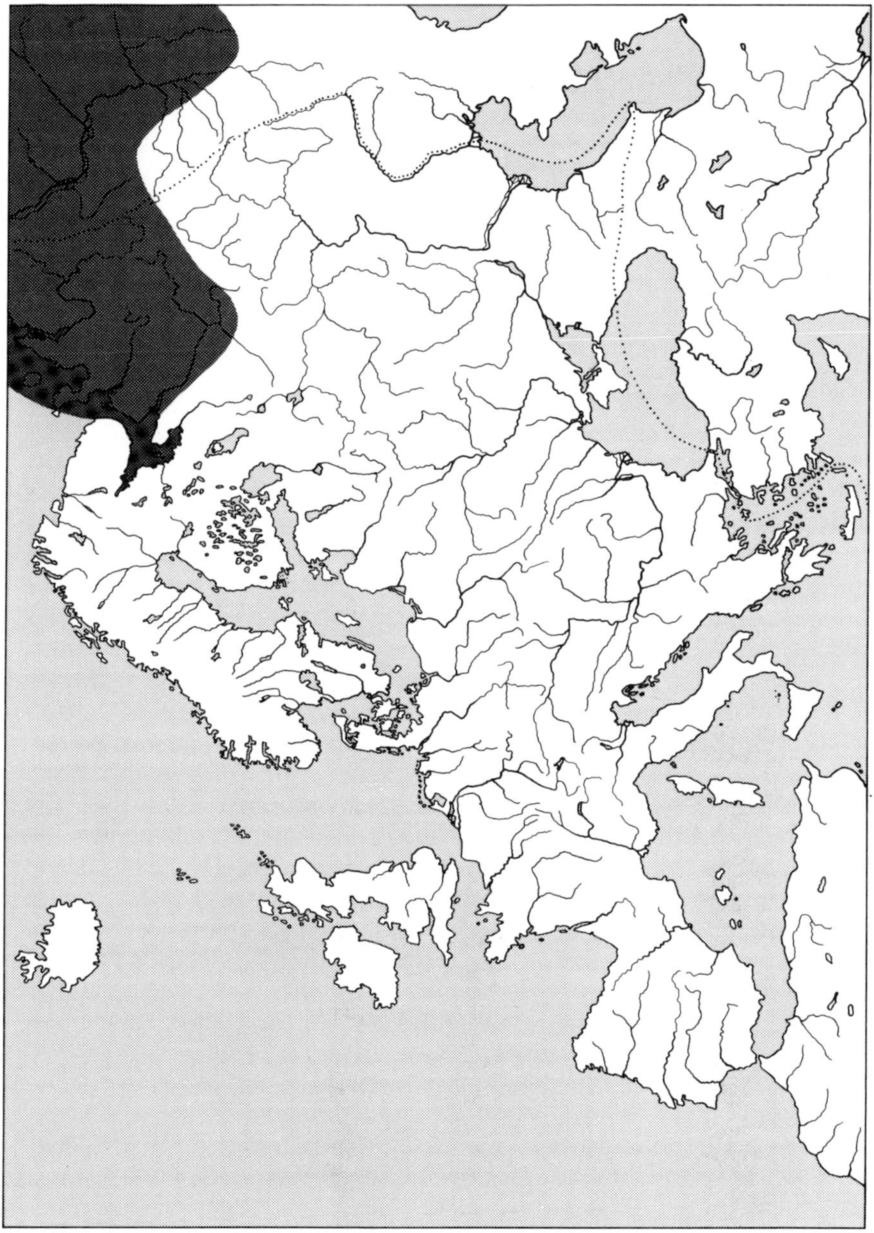

Fig. 25: European distribution of *Stenodus leucichthys nelma* (Pallas, 1773).

3.11. Osmeridae

Osmerus eperlanus (Linnaeus, 1758)

E. Smelt
F. Eperlan
G. Stint

Vernacular name: –

Family: Osmeridae

Classification: vulnerable

Note: Most subspecies and forms listed here inhabit freshwater, some of them are almost land-locked.
Osmerus eperlanus f. *spirinchus* (Pallas, 1811). This is a small form, up to 10 cm in length, with large eyes. It inhabits northern European lakes, see Kaj (1962).
Osmerus eperlanus dentex (Steindacher, 1870). This is an anadromous subspecies from the White Sea area and estuaries of Dvina and Pechora Rivers. A non-migratory freshwater population also occurs. It differs from other fishes of this group by more developed dentation. Meristic data: D 10–13, A 14–18, 65 to 72 scales on the body side.

Distinguishing characteristics

Body long and slim, head pointed and scaleless, adipose fin near the tail. The jaws are large, extending well past the level of the rear edge of the eye. The teeth in the lower jaw are large, those in the upper jaw are smaller. Large teeth are present on the roof of the mouth and on the tongue. The dorsum is light olive green; the sides have a fairly pronounced silvery stripe running from head to tail; scales small, very thin and feebly embeded in the skin. The lateral line is not complete; it is comprised of 6 to 11 scales and stops above the middle of the pectoral fin. The living animals are almost transparent, and emit a strong smell of cucumber, particularly when molested. Meristic data: D 9 to 12, A 13 to 17, 61 to 69 scales in the longest lateral row.

Worldwide distribution

Smelts are circumpolar fishes restricted to the northern hemisphere. They may be marine, anadromous or freshwater inhabitants.

Distribution in Europe

Adult anadromous fishes are rarely found far from shore and occur from the northwestern coast of Spain along the shore of the Atlantic Sea to Scandinavia. In Norway they are found in the southwest as far as Oslo-Bay. It is found along the British coastline south of Scotland, the West Coast of Ireland, and along the whole coastline of the North and Baltic Seas, including the Gulf of Bothnia.

Ecology

Adults appear to congregate near the river mouths in winter and enter the rivers in early spring (February to April). Spawning occurs during this period, or immediately after, in slightly brackish water of estuaries. The yellow eggs are attached to submerged plants. After spawning, the adults return to the sea, but the young stay in the estuary for the remainder of the summer. The smelt is an active predator, eating a wide variety of fishes, and, when small, invertebrates.

Population in Europe

Particular attention is being paid to freshwater populations of smelt. They occur in Jutland, southern Norway, Sweden, Finland and Poland. A single population in Rostherne Mere, Cheshire (United Kingdom) seems to have become extinct since 1920. For detailed informations about the population in northern Europe see Mohr (1941), Lillelund (1961) and Gaumert (1981).

Reasons for decline

Pollution of the coastal areas may be responsible for the decline, as was the case in the Baltic near Kiel, in the north of the Federal Republic of Germany. There is no up-to-date information on changes in the populations.

Conservation measures taken and proposed

Not known.

3.12. Cyprinidae

Cyprinus carpio Linnaeus, 1758

E. Common carp
F. Carpe
G. Karpfen

Vernacular name: –

Family: Cyprinidae

Classification: vulnerable to endangered

Note: The *wild* and/or *feral* carp is considered here, not the cultivated or domesticated pond forms derived from the same species.

Distinguishing characteristics

As the carp gave the name to the whole family Cyprinidae, it seems essential to mention some typical characteristics of this family. The cyprinids have a single dorsal fin and cycloid scales. The jaws are toothless; pharyngeal grinding teeth are situated under the gill cover behind the gill arches. The swim bladder is large, divided into two lobes and communicates with the alimentary canal. The first four vertebrae are modified, and a chain of small bones connects the swim bladder with the organ of hearing, the Weberian apparatus.

The wild form of the common carp has a rather elongated body, a very protrusible terminal mouth, and four barbels, two on each side of the mouth. The base of the dorsal fin is long, with a strong, serrated spine in front of the soft dorsal rays. The colour varies little. The back is usually brown to brownish green, the sides have a golden touch and the belly is yellowish. The dorsal fin is dark, the other fins have an orange tinge. Meristic features: D 20–26, A 8, 37 to 39 scales in the lateral line.

Distribution

This species is indigenous to temperate Asia from the Amur River to Yunnan including the area of the Black, Azov, Caspian an Aral Seas. Its spread into western Europe occurred via the Danube and was related to cultural events just before and after the birth of Christ. Its introduction to the American continent took place during the first half of the 19th century (see distribution map, p. 133). Domesticated forms of the carp are now present on all continents.

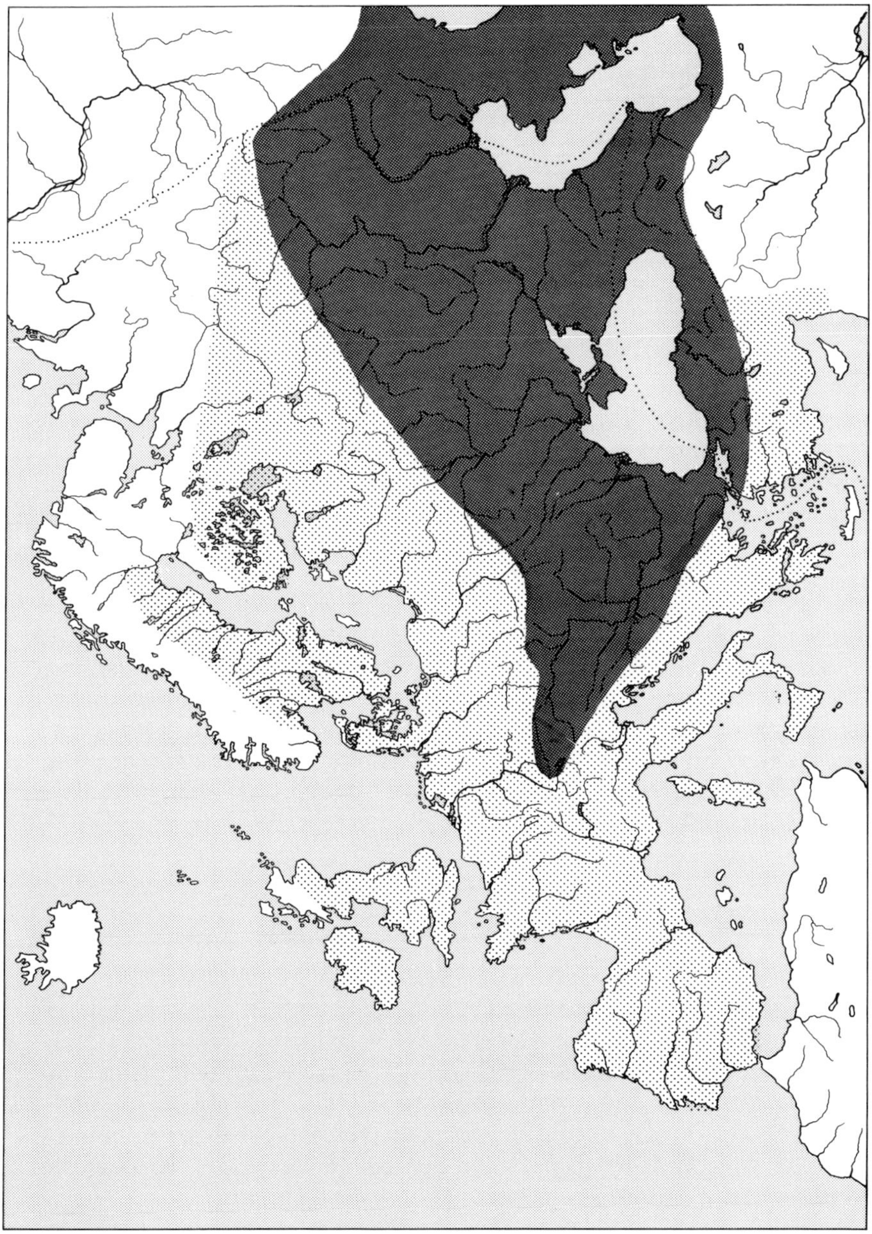

Fig. 26: Distribution of the wild form of the carp *(Cyprinus carpio)* (dark area) and the domesticated form (dotted area) in Europe and adjacent areas.

Ecology

The wild form of the common carp prefers slow-flowing streams, particularly those in which floods occur regularly. Males mature usually by the fourth year of life, exceptionally by the third. Females require an additional year for mature. Spawning occurs during May and June in flooded areas adjacent to the main rivers. A prerequisite for spawning is the vegetation, which must be fresh, preferably grass flooded with from 30 to 40 cm of water at temperatures ranging from 15° C to 19° C. Spawning groups are usually composed of one female, joined by 7 to 15 males. The eggs are released gradually, in batches of 3 or 4, within a period of 3 or 4 days, if the weather is good, or of 2 to 3 weeks if spawning is interrupted by bold, cloudy or windy spells. The eggs are attached to the vegetation and hatch from 2.5 to 3 days later. The period of planktonic feeding is short; juveniles take invertebrate food from the bottom after reaching a length of 2 cm.

A good environment is provided by oxbows and pools which are periodically joined to the main stream especially during the spring floods.

Population in Europe

It may seem unbelievable, but the wild form of the carp in Europe is in great danger. Stable, self-sustaining stocks can be found along the Danube. In the catchment area of the Rhine, the spawning places have become extremely rare. Very often the scaly, high-bodied, domesticated form is considered a substitute for the wild one. In the delta area of the Rhône, wild carp were observed in 1967. So far it was not decided whether the Rhine populations are only the feral ones.

Reasons for decline

Changes along the lower reaches of watercourses are responsible for the decline of the common carp. Straightening of the rivers, construction of high water dams and control of floodplain vegetation by draining are especially detrimental. The decline is indirectly responsible for enlarged stocking with the domesticated form. It is often falsely believed that the high-bodied, scaly carp are fast-growing wild ones. The actual lack or absence of the indigenous form is not considered because the feral and the domesticated forms are not treated separately in the management.

Conservation measures taken

At present, stocking with the cultured domestic carp, which differs to the wild form in several features is considered as a conservation measure. Actually, stocking is only a "put-and-take" procedure (when sizeable fishes are released), or a delayed "put-and-take" method (when young or undersized individuals are released). This method may be applicable in ponds set aside for angling, but it helps conservation very little.

Conservation measures proposed

Protection of the wild carp must start with the conservation of habitats suitable for natural spawning. Connection between main streams and adjacent spawning places must be guaranteed. The slope in the flooded area should be directed towards the main stream or a permanent pool, which is, at least seasonally, connected with the whole water system. The size limit for catch must be at least 45 cm total length, where wild carp occur, as the female may reach maturity at this size. The spawning places must be protected from human visitors and possible poachers, as the spawning individuals do not respond to possible danger. Monitoring of pH seems inevitable, as this gives warning of high alkalinity, which may be lethal during early stages of juvenile development.

Abramis ballerus (Linnaeus, 1758)

E. Blue bream
F. –
G. Zope

Vernacular name: –

Family: Cyprinidae

Classification: vulnerable

Distinguishing characteristics

The habitus and general features resemble the bream. The body is high and strongly compressed laterally. This species can easily be distinguished from other "breams" by the extremely long anal fin and the pointed snout, with a slightly supraterminal position of the mouth. The colour is mostly silver, turning blue-green towards the dorsum. The fins are never pink, and their edges are dark. Meristic data: D 11–12, A 38–48; 62 to 77 scales in the lateral line; 17 to 18 scales from the lateral line to the most dorsal point.

Distribution in Europe

In northern Europe, in the lower reaches of streams leading to the eastern North and the Baltic Seas, as far as the Neva; Scandinavia, only in southern Sweden and Finland; in the Danube catchment area and eastward to the Don, Volga and Ural Rivers. See distribution map, p. 137.

Ecology

This species usually inhabits sections of rivers where the current is slow, and with adjacent pools which are flooded at least seasonally. Maturity is reached in the third or fourth year of life, it spawns in places where underwater plants and a clean bottom with medium-sized gravel are available. It feeds mainly on plankton, so it was able to spread to many man-made lakes, particularly in the European part of the USSR.

Population in Europe

It is common in the Danube, downstream from Austria and is very abundant locally in Slovakia. It has almost disappeared, however, from Central European Danube tributaries such as the March River. In the northern part of its distribution area, it is not very abundant.

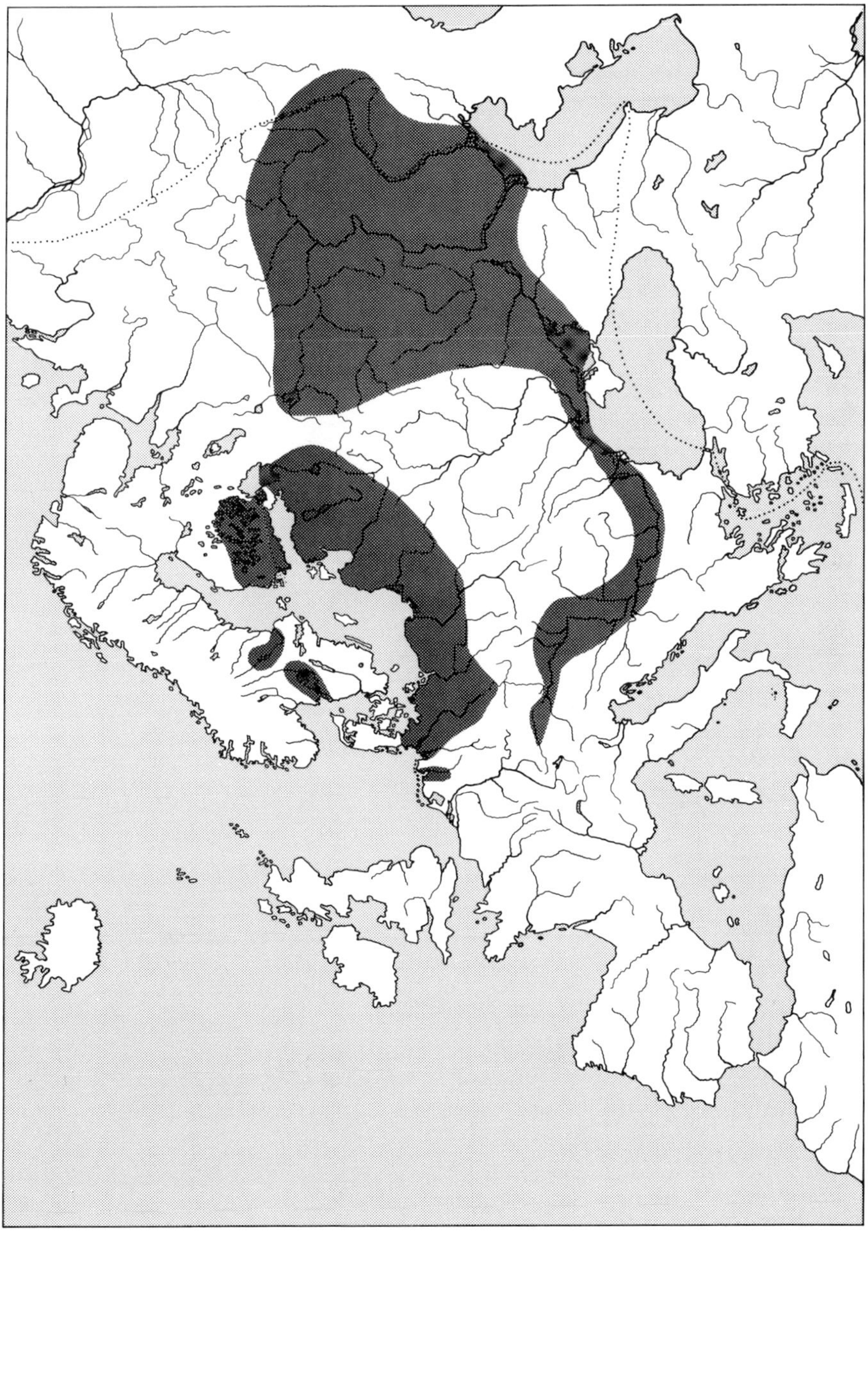

Fig. 27: Distribution of *Abramis ballerus* (Linnaeus, 1758).

Reasons for decline

Though there are no detailed studies linking its decline with poor water quality, its complete disappearance from Moravian streams may be considered indicative of this.

Conservation measures taken and proposed

No conservation measures have been taken up to now. Protection of the lower reaches of rivers within its distribution area from anthropogenic changes may be helpful.

Abramis sapa (Pallas, 1811)

E. White-eyed bream
F. –
G. Zobel

Vernacular name: –
Family: Cyprinidae

Classification: rare to vulnerable

Distinguishing characteristics

The general body habitus closely resembles that of *Abramis ballerus*. The rather blunt snout of this species, and the terminal to subterminal mouth are the main features to distinguish *A. sapa* from other species of this genus. It is darker than *A. ballerus* and has a greenish colouration, without bluish tingles. Meristic data: D 11, A 41 to 48; larger scales and consequently lower counts (48 to 52) in the lateral line; and (9) 11 (13) scales above the lateral line.

Distribution in Europe

In general, this species occurs in the Danube catchment area and rivers flowing into the Black, Caspian and Aral Seas from the north. See distribution map.

Ecology

This species lives almost exclusively in the Danube and in tributaries permanently connected with it. It spawns in the main stream, on the river bottom which must be free from sediment or detritus. Maturity is reached in the fourth year; the growth rate is the slowest of all breams mentioned. The diet consists mainly of benthic fauna. In the eastern part of its distribution area, spawning migrations from the brackish waters into the streams have been observed.

Population in Europe

This species has never been abundant. Because its sport and commercial value is negligible, little is known about its populations in European waters.

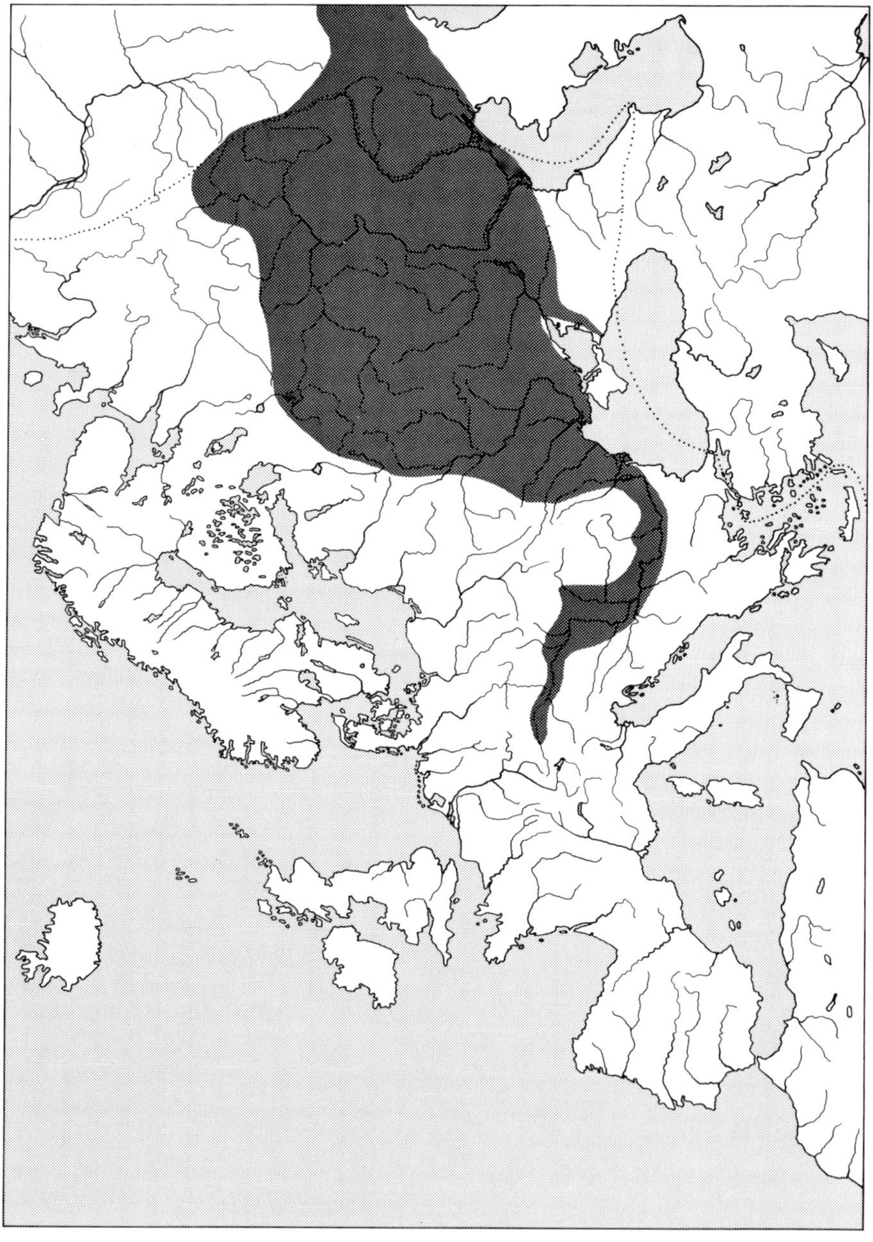

Fig. 28: European distribution of *Abramis sapa* (Pallas, 1811).

Reasons for decline

Its disappearance from Danube tributaries in Moravia probably proves its sensitivity to poor water quality. The heavy ship traffic may possibly be responsible for its recent decrease in abundance.

Conservation measures taken and proposed

No conservation measures are known and none have been proposed. It is believed that this species will be adversely affected by proposed construction projects on the Danube. As it is already rare, it may disappear completely.

Alburnus albidus (Costa, 1838)

E. –
F. –
G. –

Vernacular name: Alborella (Italian)

Family: Cyprinidae

Classification: intermediate

Note: In this species are inlcuded three rather dubious subspecies: *Alburnus albidus alborella* (De Filippi, 1844), found in the rivers of northern Italy, Lakes Skutari and Ohrid, and rivers in Albania; *Alburnus albidus belvica* (Karaman, 1928), found in Lake Prespa and *Alburnus albidus scoranzoides* (Heckel and Kner, 1858). (Often *Alburnus alburnus albidus* is treated as a subspecies of *Alburnus alburnus.*)

Distinguishing characteristics

The body habitus of this species is similar to that of the bleak and it reaches the same small maximum size of 20 cm. Of the meristic features the most important is the anal fin: A 14 to 16 rays which make it distinguishable from the bleak; D 10 to 12. Pharyngeal teeth in two rows.

Distribution

In southern Europe, only Italy and the Balkan Peninsula. See distribution map.

Population

Not much is known about this fish, but it is assumed that *Alburnus albidus* replaces *Alburnus alburnus* within its distribution area.

Ecology

Similar to that of the bleak, *Alburnus alburnus.*

Conservation measures taken or proposed

No conservation measures known, none proposed.

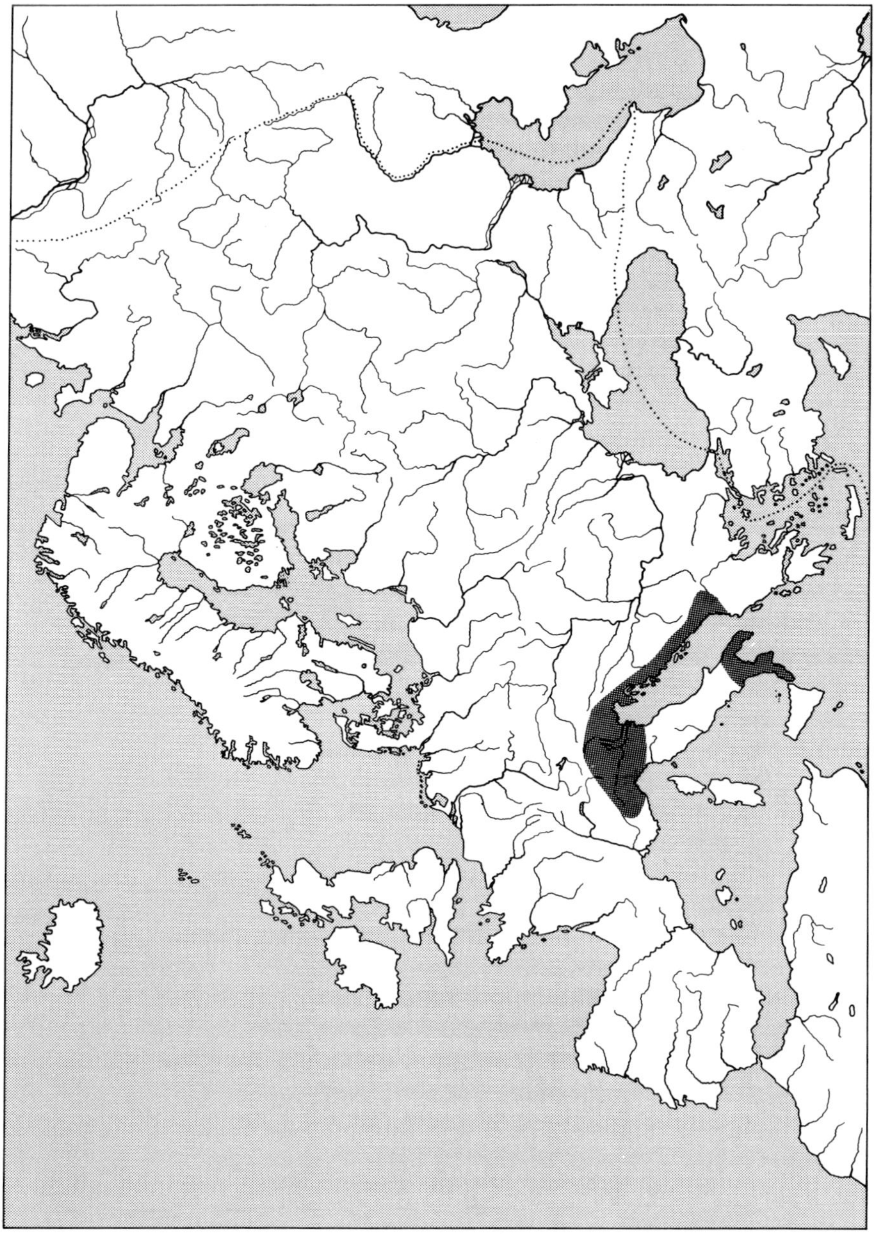

Fig. 29: Area of distribution of *Abramis albidus* (Costa, 1838).

Alburnus alburnus (Linnaeus, 1758)

E. Bleak
F. Ablette
G. Ukelei

Vernacular name: –

Family: Cyprinidae

Classification: intermediate

Within a large distribution area, two sub-species have been described: *Alburnus alburnus macedonicus* (Karaman, 1928) form the Wardar River and *Alburnus alburnus strumicae* (Karaman, 1955) from the Struma and Maritza catchment areas. For detailed informations on local races see BĂNĂRESCU (1964).

Distinguishing characteristics

This species is a slender silvery fish with a maximum length of 18 to 20 cm. The mouth is superior and the lower jaw prominent, but the corners of the mouth do not extend beyond the front edge of the eye. The anal fin is long, its first ray directly beneath the last three dorsal rays. There is a sharp, scaleless keel between the pelvic and the anal fins. Colouration: the dorsum is grey-green, the sides and particularly the belly brilliant silver. Meristic data: D 10 to 12, A 17 to 23, pharyngeal teeth in two rows, usually 2 plus 5 each side, slightly rigid.

Distribution in Europe

The bleak is one of the most common European fishes. It is found in the eastern part of England, but not in Ireland or Scotland. Eastwards it is distributed throughout Scandinavia, with the exception of western part of Norway, in those rivers flowing to the Atlantic, and from the Kola Peninsula to the Ural Mountains. Further it occurs from the northern Pyrenees throughout Europe except the Po and Adige catchment areas in Italy, Corsica, Sardinia and Sicily. In the Black Sea region it is not found in the southern catchment area and in the area of the Caspian Sea it occurs only in rivers emptying in the northern part of the sea. A relatively small catchment area, that of the Kuban and Terek Rivers, is occupied by another *Alburnus* species. In the east the range ends with the catchment areas of the Volga and Ural Rivers.

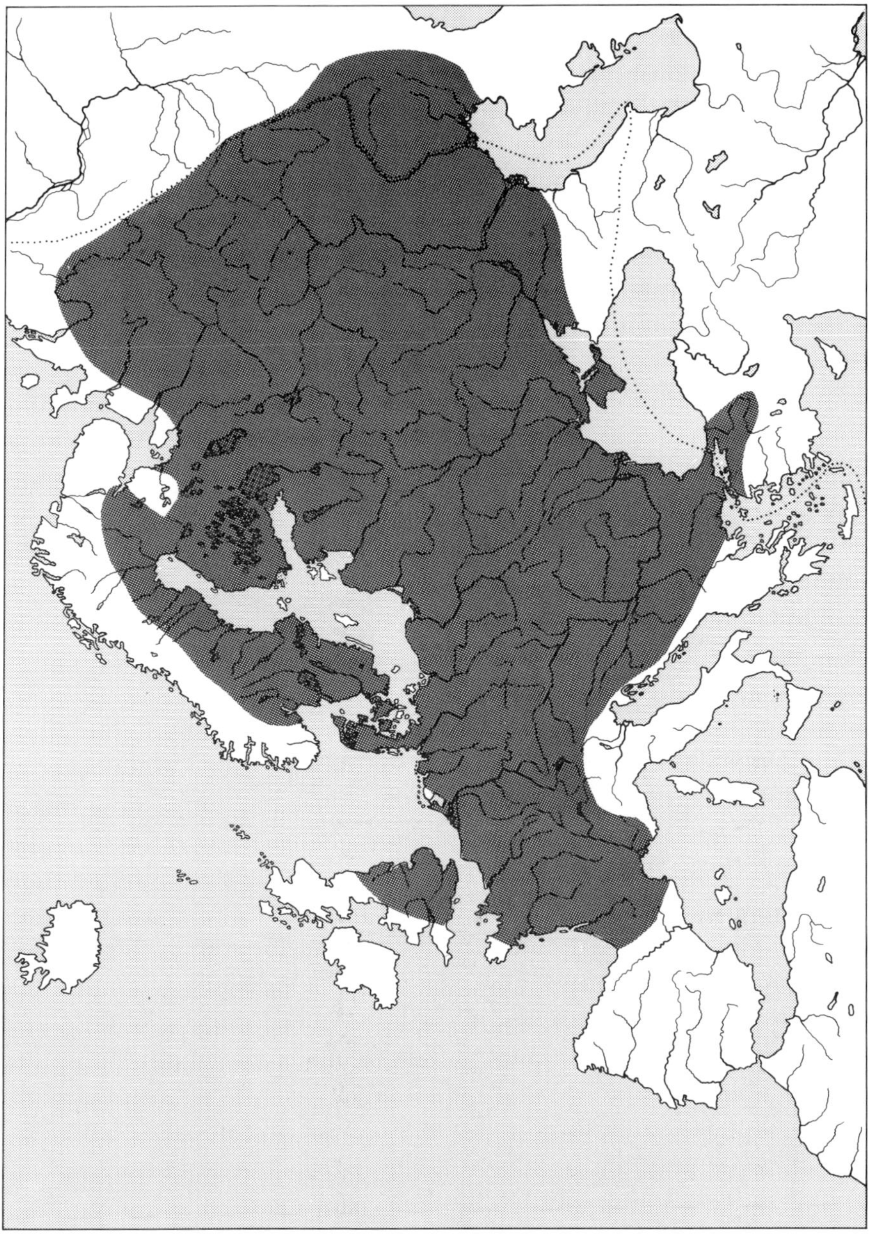

Fig. 30: Distribution of *Alburnus alburnus* (Linnaeus, 1758).

Ecology

The bleak lives in running, slow-flowing and lentic environments, preferring larger slow rivers; it is a gregarious pelagic fish, always found near the surface. It does not need water fully saturated with oxygen and can withstand some pollution caused by domestic wastes. It spawns rather early, from mid-April to May. The eggs are released in batches over a fairly lengthy period. It does not need specific spawning substrate, and can utilise grasses, roots, stony bottom, flotsam, vegetation remaining from the previous year, stony shores or concrete. It feeds on the eggs and larvae of earlier spawners; when chironomids emerge or terrestrial prey is blown on the water, it feeds preferably from the water surface. It does not live long; males mature in the 2nd and females in the 3rd year and the oldest individuals attain only 5 to 6 years.

Population in Europe

Now one of the most common species in some European waters, it has no special spawning requirements or habitat, and few natural enemies. Locally, it is fished commercially and used for human consumption.

Reasons for decline

Decline has been observed only locally, particularly after contamination with lethal pollution. As this species is rather vagile, it repopulates habitats rapidly. Lasting pollution may be a danger to populations in some areas.

Conservation measures taken and proposed

No conservation measures known. None required at present.

Alburnoides bipunctatus (Bloch, 1785)

E. Spirlin
F. Spirlin
G. Schneider

Vernacular name: –

Family: Cyprinidae

Classification: vulnerable to endangered

Note: Several subspecies have been described, some of which occur in the European region: *Alburnoides bipunctatus rossicus* (Berg), which is distributed in the Volga catchment area; *Alburnoides bipunctatus strymonicus* (Chichkoff), in the catchment area of the Maritsa River; *Alburnoides bipunctatus ohridanus* (Karaman) and *Alburnoides bipunctatus fasciatus* (Nordmann), in Lake Ohrid and the Drin and *Alburnoides bipunctatus prespensis* (Karaman) from Lake Prespa. 6 other subspecies occur in the Asian watersheds of the Black and Caspian seas.

Distinguishing characteristics

The spirlin is a small silvery cyprinid fish reaching a maximum length of 13 to 14 cm. The mouth is terminal and extends rearwards as far as the eye. The colouration is rather conspicuous. The dorsum is olive green; the sides silver with a yellowish shine along the clearly visible, ventrally bent lateral line, which looks like a row or paired points. Meristic data: D 10 to 12, A 16 to 20; 44 to 51 scales in the lateral line.

Distribution in Europe

The European range extends from France eastwards to the rivers feeding the Caspian Sea. It does not occur south of the Alps and Pyrenees, nor in Scandinavia. See distribution map, p. 148.

Ecology

The spirlin is found in small shools, usually in medium sized streams, but less frequently in the main channel of rivers. It lives in unpolluted, well-oxygenated stream sections, always in currents, and it can tolerate higher temperatures than brown trout and grayling. Preferred habitats are those with well modified, hard, stony or rocky bottoms. Spawning takes place in late spring. The eggs are deposited on sand or gravel in swiftly flowing

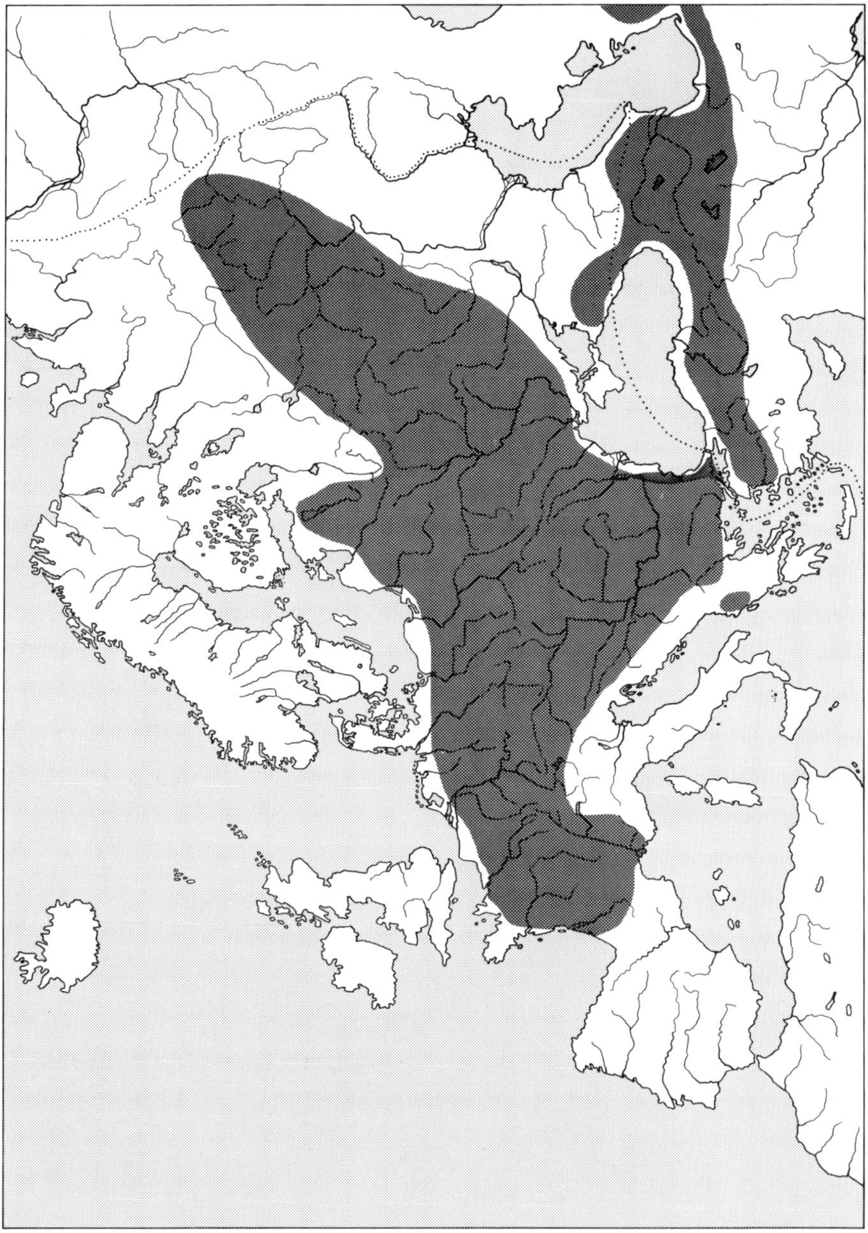

Fig. 31: Distribution of *Alburnoides bipunctatus* (Bloch, 1758).

water. The food is taken both from the bottom (mostly insect larvae) and from the surface (terrestrial drift).

Population in Europe

It has been observed to be a very abundant species in medium-sized streams of the Danube area and elsewhere. In the Elbe and Rhine catchment areas, abundance was always low. During the last few decades, its populations have been much reduced. For instance, it is extremely rare in tributaries of the Rhine, it disappeared from the Main River completely. In the Danube from its source downstream to Austria, it is extremely rare in the main stream. As it has no value, economically or for sport, little is known of its occurrence.

Reasons for decline

It is believed that organic pollution and general eutrophication of water courses have caused the decrease in numbers.

Conservation measures taken

None known, except legal protection in some countries.

Aspius aspius (Linnaeus, 1758)

E. Asp
F. –
G. Rapfen

Vernacular name: –
Family: Cyprinidae

Classification: vulnerable to endangered

Note: Two subspecies of *Aspius aspius* occur outside Europe: *Aspius a. taeniatus* (Eichwald, 1873) in the southern Caspian Sea and rivers feeding it; *Aspius a. ibloides* (Kessler, 1872) in the catchment of the Aral Sea, and a related species *Aspius vorax* (Heckel, 1843) in the basin of the Euphrates River.

Distinguishing characteristics

The asp is a slim fish with an elongated, compressed body and large mouth, reaching to at least the posterior third of the eye. The lower jaw projects, and its tip fits into a notch in the upper jaw. The anal fin is relatively long and bright; its first rays are long and the whole fin is conspicuously concave. The fish may weigh over 8 kg. The dorsum is dark grey-green, the sides silvery, but not shiny. Meristic data: D 11 to 12, A 15 to 18 and 64 to 76 scales in lateral line.

Distribution in Europe

The distribution extends from the Elbe eastward to the Ural Mountains. It occurs in Scandinavia eastwards from Oslo, in southern Sweden and southern Finland. The catchment area of the Danube forms the southwestern limit of its range. It was absent from the Rhine and further westward, including the British Isles. Solitary individuals were ascertained in the Rhine and Main Rivers since 1985 as a result of stocking from the Danube catchment area. In south-eastern Europe, it occurs in the Struma and Maritsa Rivers. See distribution map.

Ecology

The asp usually inhabits running waters, although it has been found in larger lakes. After reaching 20 to 30 cm, it preys on fish. The young fishes are gregarious. Older ones live solitary, but occasionally feeding in schools of individuals of more or less the same size, which swim close together

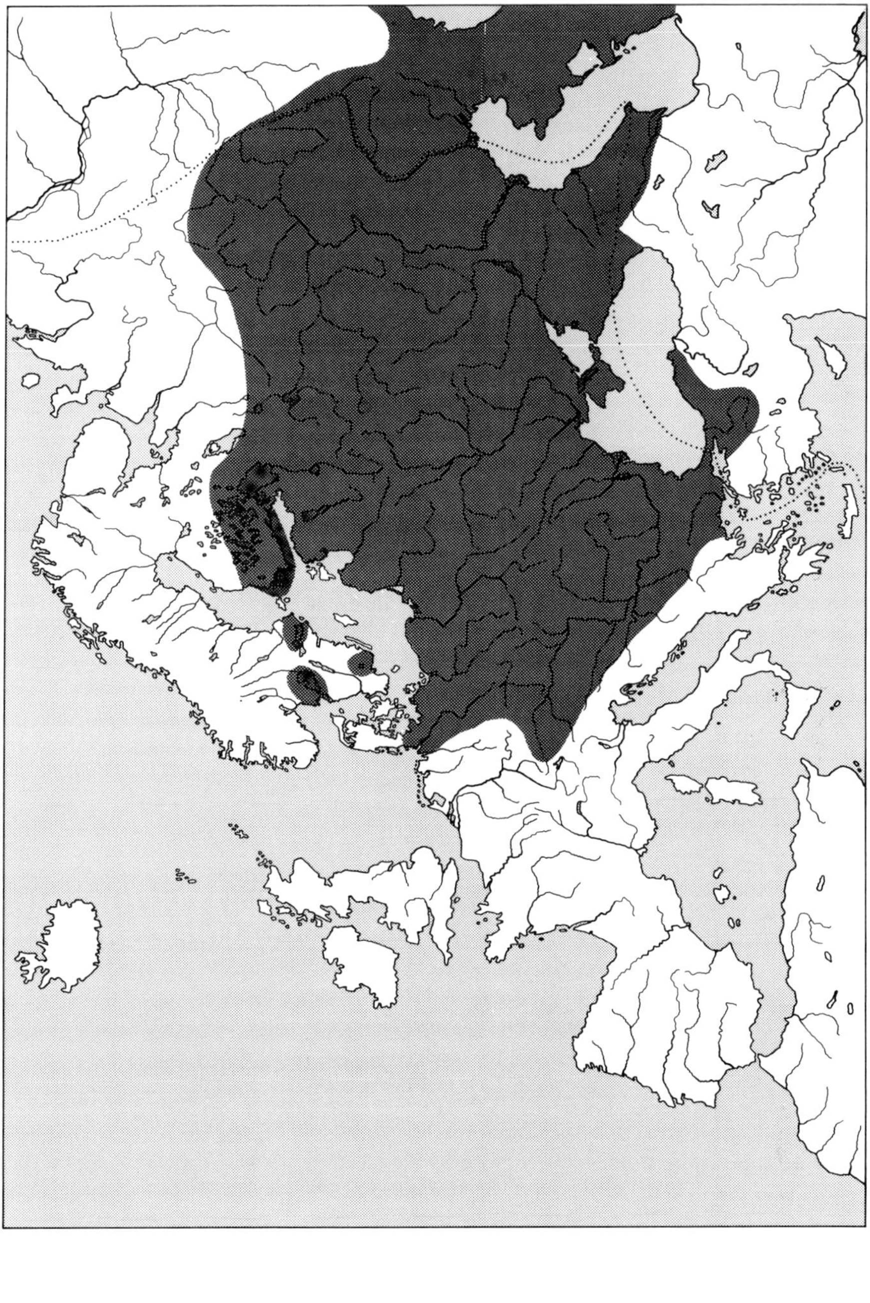

Fig. 32: European distribution of *Aspius aspius* (Linnaeus, 1758).

seeking their prey. Maturity is reached between 4 to 5 years, males being mature one year earlier than females. It spawns in April or May, in running water above a clean sandy or pebbly substrate. The mainly pelagic larvae drift downstream into more lentic sections. Growth is rather fast. A length of 10 to 20 cm is reached in the first year and 30 to 47 cm by the third year of life. Ten-year-old individuals may reach 60 cm and weigh 2.5 kg. Individuals longer than 80 cm are rare, but some were seen in 1976 in German and Austrian stretches of the Danube. The life-history of this species is not known in detail.

Population in Europe

In general, except some little changed stretches of the Danube the European population seems endangered. It is rarer than it was several decades ago. In Scandinavian waters, it is definitely endangered.

Reasons for decline

Environmental changes, particularly the damming of watercourses, have affected this species in the Danube. It occurs only in habitats where lotic conditions prevail. Silting caused by impoundments probably destroyed many spawning places. It is believed that slight deterioration of water quality is not in itself detrimental to this species.

Conservation measures taken

Legal measures have been taken in almost all countries (size limit and closed seasons), but this does not seem sufficient.

Conservation measures proposed

Learning more of the life-cycle may help in finding methods for practical conservation. At present, only some generally useful activities, as listed in Chapter 4, can be suggested. Restocking of rivers within its former distribution area is suggested, as well as experimental work to find ways of artificial propagation of the species.

Aulopyge hügeli Heckel, 1841

E. –

Vernacular name: Ostrulj (Croatian)

F. –

G. Barbengründling

Family: Cyprinidae

Classification: rare

Distinguishing characteristics

This species is found only in Dalmatia and Bosnia, and it is difficult to compare it with other taxa. It has a streamlined body. The lateral line is complete but does not run straight along the side. The body is scaleless. The mouth is ventral with four barbels; the snout is pointed and narrow. The body is silvery, with irregular dark blotchs and spots; the dorsum is green. There is a pronounced sexual dimorphism, females being more corpulent and having a higher body than males. Both sexes reach a length of 13 cm. Meristic data: D 10, A 7.

Distribution and population in Europe

It is native to Dalmatia, and is found only near Livno in Bosnia, in waters of the Dinaric Karst area. See distribution map, p. 154.

Ecology

The ecology of this species is not known in detail.

Reasons for decline

No data are available for comparing past and present populations.

Conservation measures taken and proposed

No conservative measures are known, but it is strongly recommended that the environment of this species should be protected. It is known that relict species are very sensitive to environmental changes. This endemic species may be compared to *Romanichthys valsanicola* in Romania (see p. 291). It is hoped this taxon will not meet the same fate.

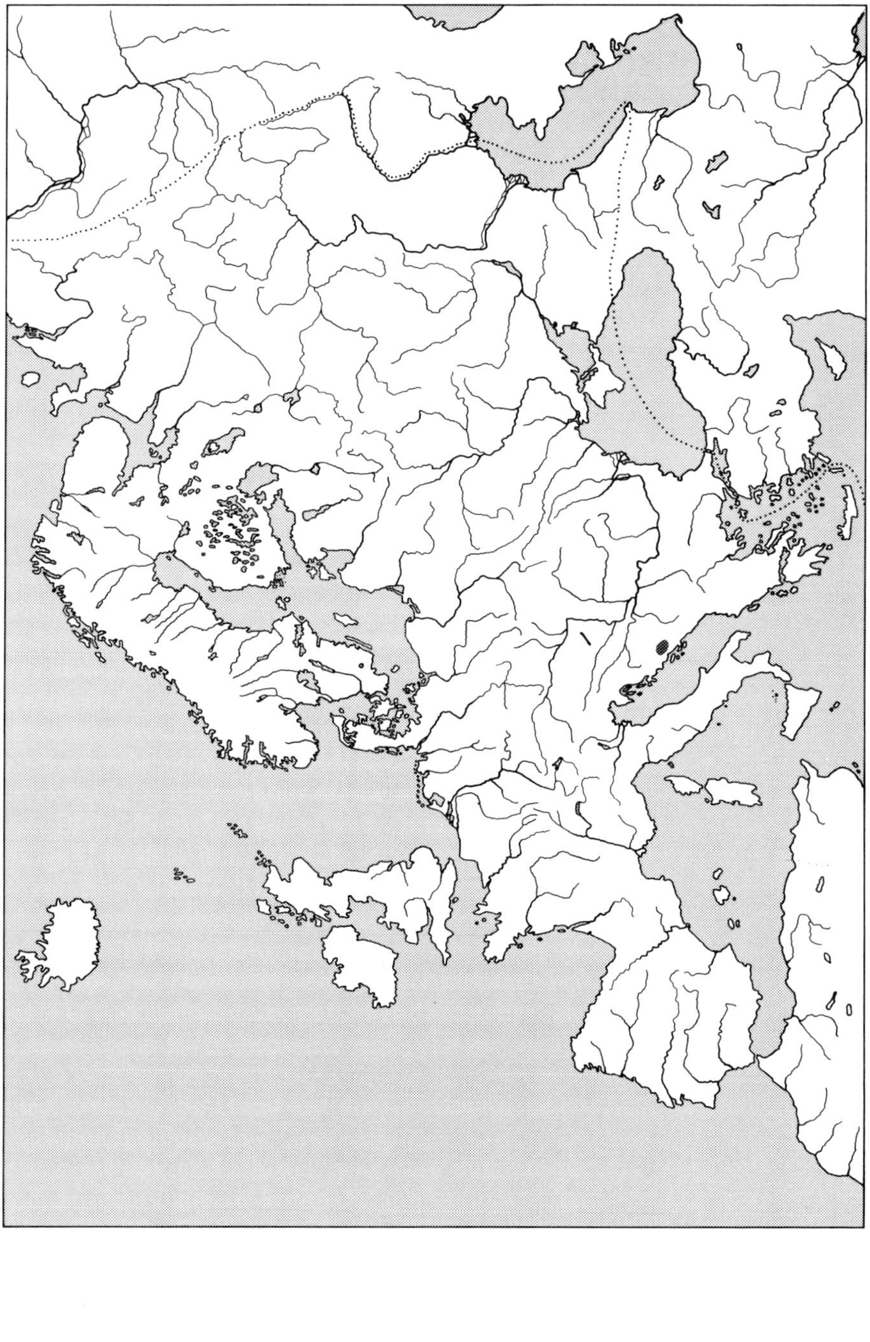

Fig. 33: Distribution area of the to Dalmatia endemic species *Aulopyge hügeli* Heckel, 1841.

General notes to the genus *Barbus*

The genus *Barbus* is another category requiring additional systematic and taxonomic investigation to provide a sound basis for a valid nomenclature. Most species of this genus are non-migratory and confined to one watershed or to one particular, often small part of a catchment area. Because of the isolation of many species, caused by hydrographical changes in Europe during the past, certain characteristic features have developed. The richness of species and their sub-specific categories is observed in streams of the Balkan Peninsula, eg. *Barbus peloponnensius* Valenciennes, 1842 and in south-western Europe, eg. *Barbus comiza* (Steindachner, 1866) which is accompanied by three other species of *Barbus* (see map 36, p. 164). Hopefully the increasing environmental awareness will help to secure the habitats from irreparable changes. At present, there is no evidence that the *Barbus* species of the European South, though vulnerable to anthropogenic impacts, are seriously endangered. Moreover, hybridization among forms occurs easily (e.g. *B. barbus* x *B. meridionalis*) which makes further classification difficult. It is hoped that the problems of taxonomy and synonomy will be definitely resolved and criteria established for species and sub-species on the basis of conventional morphology, using readily available statistical methods. This may be easier to solve than the problem of salmonids or coregonids, where massive transfers of foreign forms from one water body to another has caused loss of genetic identity (introgression). In the case of the barbel, no attempt to stock new water bodies artificially with foreign forms has been reported. Some repopulation experiments have been made, sometimes with success, to repopulate river stretches with barbel after its elimination by pollution. Solving difficulties in taxonomy will require additional research.

Proposals for protection and conservation, which cannot be postponed, are based on early zoogeographical results, such as those of Berg (1933) and Bănărescu (1964) as well as recent work by Karaman (1971) and Almaça (1978). In this study, however, the nomenclature suggested by Ladiges (1978) was used. It must be stressed that taxa occurring in Central Europe are, for obvious reasons, treated in more detail. Of the 38 taxonomic forms listed in Bănărescu *et al.* (1971), only three will be dealt with in this volume, as ecological informations about others are not readily available,

Barbus barbus (Linnaeus, 1758)

E. Barbel
F. Barbeau fluviatile
G. Barbe

Vernacular name: –

Family: Cyprinidae

Classification: vulnerable

According to Ladiges (1978) there are three subspecies of this species at the edges of the European distribution area: *Barbus barbus gallicus* Kar., *Barbus b. macedonicus* Kar. and *Barbus b. thessalus* Steph. More about the subspecies of the barbel will bring one of the next volumes of this series.

Distinguishing characteristics

The barbel is a long, round-bodied fish with a long head and a pointed fleshy snout. It has four barbels, two near the snout tip and two at the angle of the jaw. The third and last hard ray is serrated posteriorly. The anal fin, when pressed toward the body, does not reach the insertion of the caudal fin. The colour is variable. The dorsum is usually brown-green and the greenish-golden-yellow flanks show transition to whitish-yellow on the belly. The iris is golden. Young individuals have uniform colouring on the sides and back, and may have scattered dark-brown spots on the sides of the body. Meristic data: D 10 to 12, A 8; 56 to 65 scales in the lateral line; pharyngeal teeth in 3 rows.

Distribution in Europe

The barbel is found from France to the Neman River in the Northeast. In England, it is native only to some southeastern rivers and is now present in the catchment areas of the Severn, Avon, Thames, Welland and Trent, and possibly some others. Its eastern distribution limit is the watershed of the Dnieper. In the south, it occurs in the Rhône and in the whole of the Danube catchment area. See distribution map, p. 157.

Ecology

The barbel is a typical species of river stretches with a varying moderate gradient, a clean bottom with sand and gravel, moderate to powerful currents. These are therefore called "barbel zone". In this zone the brown

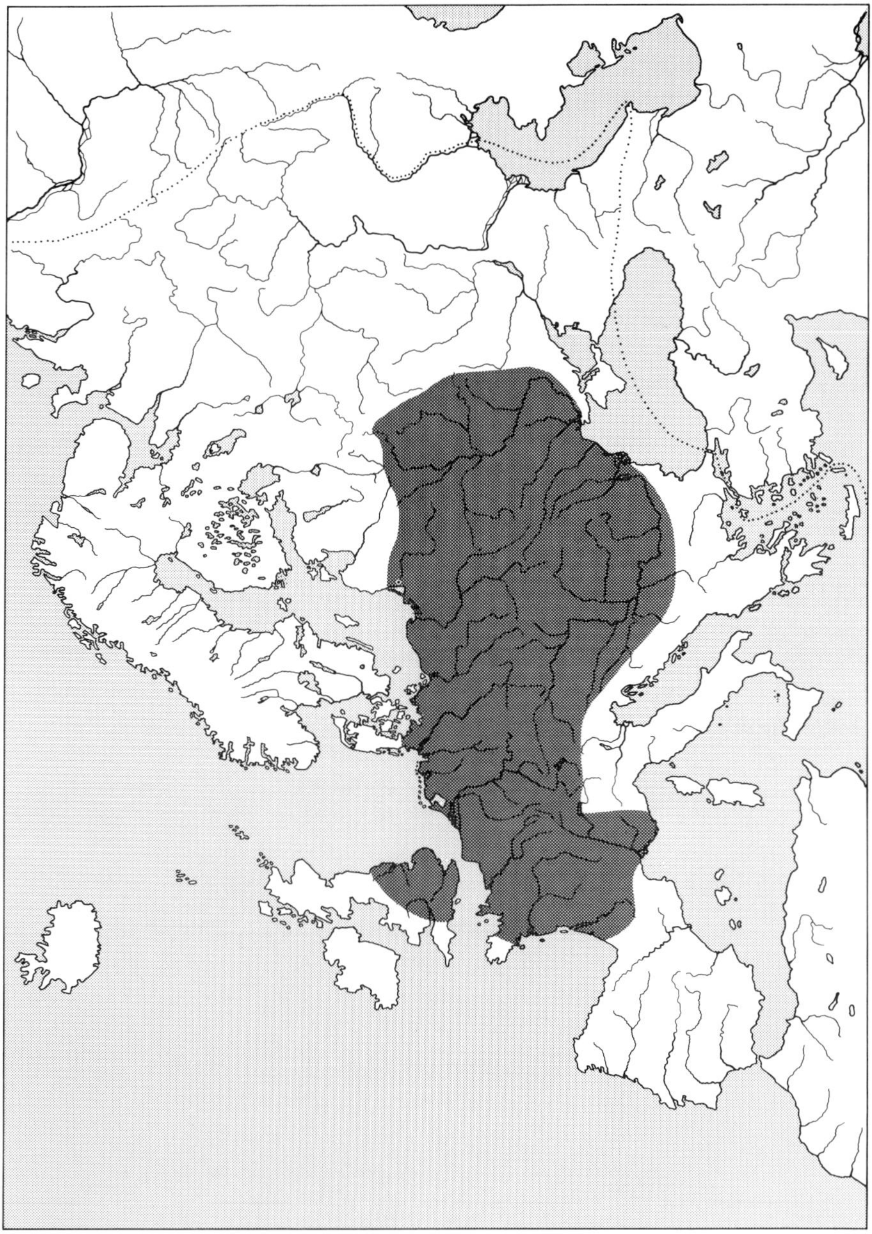

Fig. 34: Distribution of *Barbus barbus* (Linnaeus, 1758) in Europe.

trout gradually disappears and the grayling is no longer the most abundant species. The "barbel zone" is usually dominated by cyprinids. This description applies to Central European waters and particularly the Danube catchment area. The barbel is rare in lower stretches with muddy bottom or in areas with much silting. It prefers the middle of currents and it does not congregate in shoals but always occurs in groups of several individuals, the largest ones occupying deeper zones of current. In shallow, swiftly flowing water, smaller individuals are predominant. Spawning takes place in medium depths, where gravel is mixed with sand and pebbles. For spawning, it requires a clean bottom in the main stream with a diversified structure, preferably in and immediately upstream from the confluence of main stream and a tributary. The eggs are firmly attached to stones. The larvae are photophobic. In the early stages, the fish stay close to the shore line, often in shallows, but always in swiftly running water.

Sexual maturity is generally reached by the males in the fourth and the females in the fifth year of life. The growth rate is slow, from 7 cm at one year, to between 20 to 25 at five years. Individuals of 60 to 80 cm are well over 10 years old. The barbel is a typical bottom fish, seeking food among stones and pebbles with its fleshy lips. It lives on insect larvae, worms, large crustaceans and occasionally small bottom-dwelling fish. Stretches with both fast and slow currents and refuges in the centre of the stream are a prerequisite for its permanent abundance and spawning success.

Population in Europe

Provided the physical characteristics of river sections (see above) do not change populations will contain fish of all age groups. The barbel is locally abundant and can withstand moderate anthropogenic pollution. The appearance of large, solitary individuals indicates deterioration of a stable population.

Reasons for decline

Modifications of rivers, especially monotonous canalization of long stretches, causes the gradual disappearance of *Barbus barbus*. Further it is believed that the barbel is sensitive to pollution by detergents.

Conservation measures taken

In many countries, the barbel is legally protected, mainly by closed seasons and size limits.

Conservation measures proposed

As this species lives in river streches passing human agglomerations, such as small towns, villages and small industries, it is extremely difficult to improve and protect the habitat. This species is one for which graduate improvement of water quality has only limited effect. The only permanent solution may be the creation of "fish sanctuaries", as proposed in section 4, page 316.

Barbus meridionalis Risso, 1826

E. – Vernacular name: –
F. Barbeau truité
G. Semling Family: Cyprinidae

Classification: vulnerable

Note: The subspecies *Barbus m. petenyi* (Heckel, 1847) occurs in the Danube basin.

Distinguishing characteristics

The body habitus strongly resembles the barbel, *Barbus barbus,* the primary apparent difference is the colouration: *Barbus meridionalis* is generally darker. The back is dark brown, often with almost black irregular spots extending to the flanks. The belly is cream to white. Taking only the colouration into account, this species may easily be confused with young *B. barbus.* The most important distinguishing character is the lack of serration on the posterior edge of the third (longest) dorsal spine. The anal fin, when pressed to the body, reaches or very slightly overlaps the insertion of the caudal fin. Meristic data: D 11, A 8; (49) 51 to 55 (57) scales in the lateral line; pharyngeal teeth in 3 rows.

Distribution

This species occurs on the Iberian Peninsula, in part of the Rhône River catchment area, in the rivers of Northern and Central Italy, as well as farther east in the Danube catchment area. This region is clearly separated from the ranges in western Europe, its westernmost limit being the Hron and Ipel Rivers in Slovakia. The northernmost occurance of *B. meridionalis* is found in the upper parts of the Vistula River and the limits of range in the East are the tributaries of the Dniester River. See distribution map, p. 161.

Ecology

This species is a bottom-dweller with many similarities to the barbel. *B. meridionalis* occupies river stretches with the same physical conditions, but generally upstream from the habitat of the barbel. In some eastern Euro-

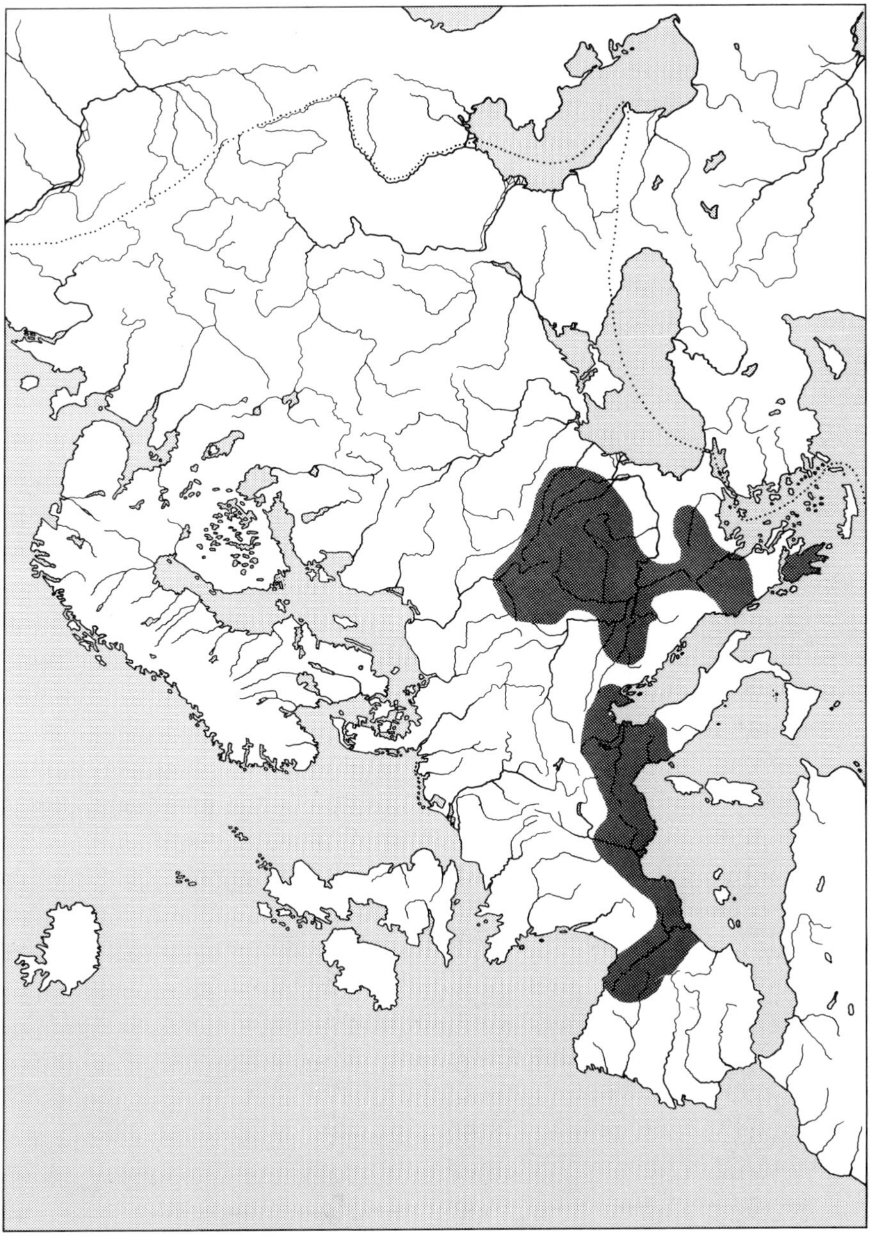

Fig. 35: Distribution of *Barbus meridionalis* Risso, 1826.

pean streams, it occupies the grayling zone, and often replaces the grayling itself. As it often shares habitat with the barbel, hybrids of both species are not unusual. Feeding habits are also the same as those of the barbel. It reaches a maximum length of 25 cm; larger individuals are rare. Little is known of its ecology. This species demands cool water, well saturated with oxygen, and it is very sensitive to all types of pollution.

Population in Europe

Where the river stretches within its range are not changed by human activities, it is abundant. It has no commercial value and is little prized as a sport fish.

Reasons for decline

Pollution or changes of currents and bottom substrate may be detrimental to the survival of this species.

Conservation measures taken and proposed

No conservation measures are known. Protection measures should begin with the environment (see section 4, page 316).

Barbus plebejus Valenciennes, 1842

E. – Vernacular name: Barbo (Italien)

F. –

G. – Family: Cyprinidae

Classification: vulnerable

Note: The following subspecies have been described and were listed by Ladiges (1978): *Barbus p. ciscaucasius* Kessl., *Barbus p. cycloepis* Heckel, *Barbus p. euboicus* Seph., *Barbus p. prespensis* Kar., *Barbus p. tauricus* Kessl., Almaça (1983) considers *B. haasi* Mertens, 1924 to be a subspecies of *B. plebejus*.

Distinguishing characteristics

The body habitus is typical of the barbels. The ventral line is straight, the dorsal side dark brown and the belly bright. All over the body are found irregular dark spots. The dorsal, anal and caudal fins are marked with small dark blotches.

Distribution in Europe

Found south of the Alps, mostly in Italy. No detailed information on populations is reported. The occurrence in Slovakia, Krupka & Holčik (1976), which is off the distribution areal, has to be reconfirmed. See distribution map, p. 164.

Ecology

Not much of its ecology is known.

Conservation measures

No comments on conservation can be made.

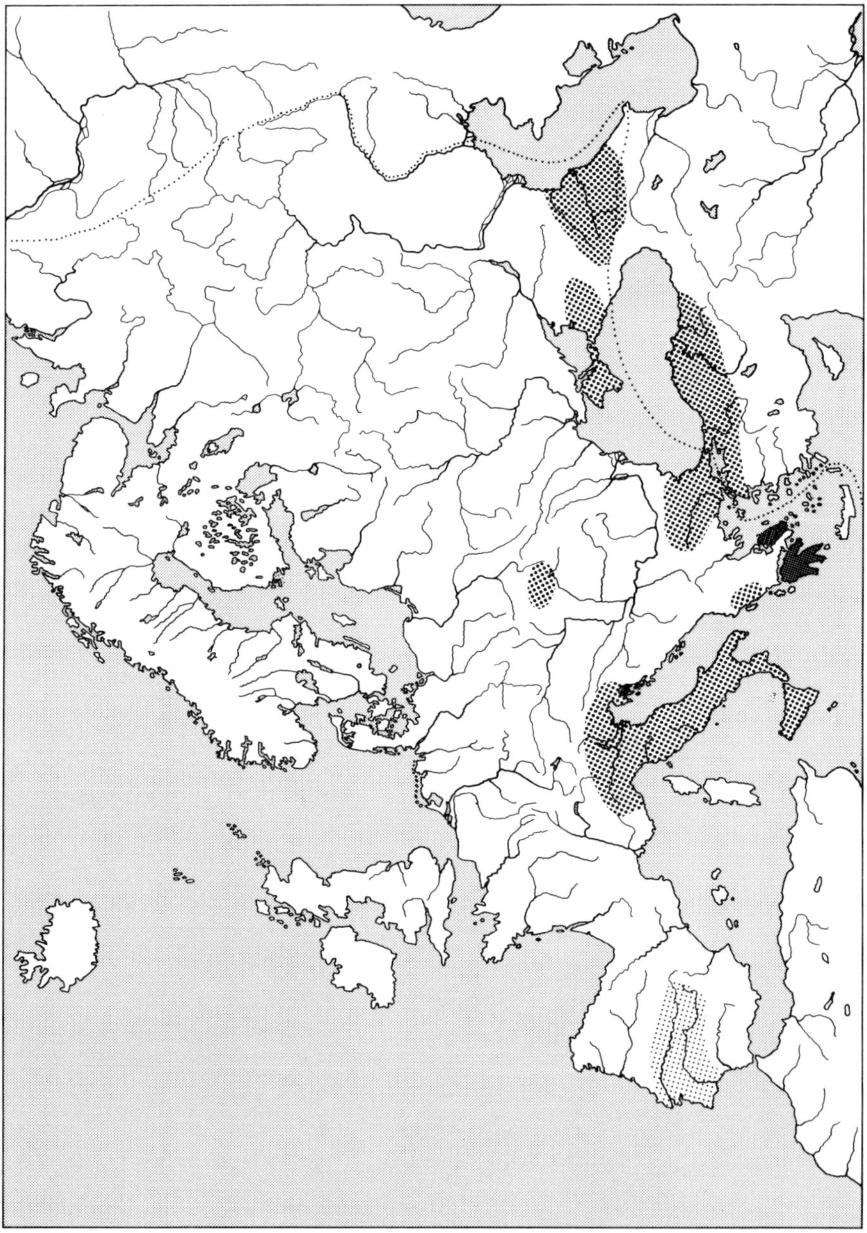

Fig. 36: Distribution of *Barbus plebejus* ▒, *Barbus comiza* ░ and *Barbus peloponnensius* ▓.

Blicca bjoerkna (Linnaeus, 1758)

E. White (Silver) bream
F. Brême bordelière
G. Blicke, Güster

Vernacular name: –

Family: Cyprinidae

Classification: intermediate

Note: The subspecies *Blicca b. transcaucasica* (Berg 1916) was ascertained in Turkey (Ladiges, 1960).

Distinguishing characteristics

This is a high-bodied fish with long anal fin, partly resembling *Abramis ballerus*, and *A. sapa,* with its moderately hump-backed body and inferior situated mouth. The eye is relatively large, and its diameter is greater than the snout length. The colour of the fins is an important diagnostic feature: the pelvic and pectoral fins are orange or reddish near the base. The number of rays is characteristic: up to 24 in the anal fin (exceptionally 26). Meristic data: D 11, A 22 to 24 (25); 44 to 48 scales in the lateral line; pharyngeal teeth in two rows.

Distribution in Europe

The white bream is found throughout northern Europe, the Pyrenees and the Alps. In England only in the southeast. It is also found in southeastern Norway, southern Sweden, and Finland. It is not present in the rivers feeding Lakes Onega and Dvina. In the east, its range includes the Volga and Ural River basins and the Caspian and Black Seas as far as the Terek and Kura Rivers. Its southwestern distribution area reaches to the north of the Pyrenees. It is rare in the Garonne, but is common throughout Central Europe. *Blicca bjoerkna* is not found in the rivers, emptying in the Mediterranean Sea. In the Southeast of Europe its range reaches to the catchment area of the Kamchiya River in Bulgaria. See distribution map, p. 166.

Ecology

White bream usually occurs in the lower reaches of the larger, slow-flowing rivers, adjacent plains, and periodically flooded waterbodies. It may be very common in some man-made lakes. Spawning time is rather late, usu-

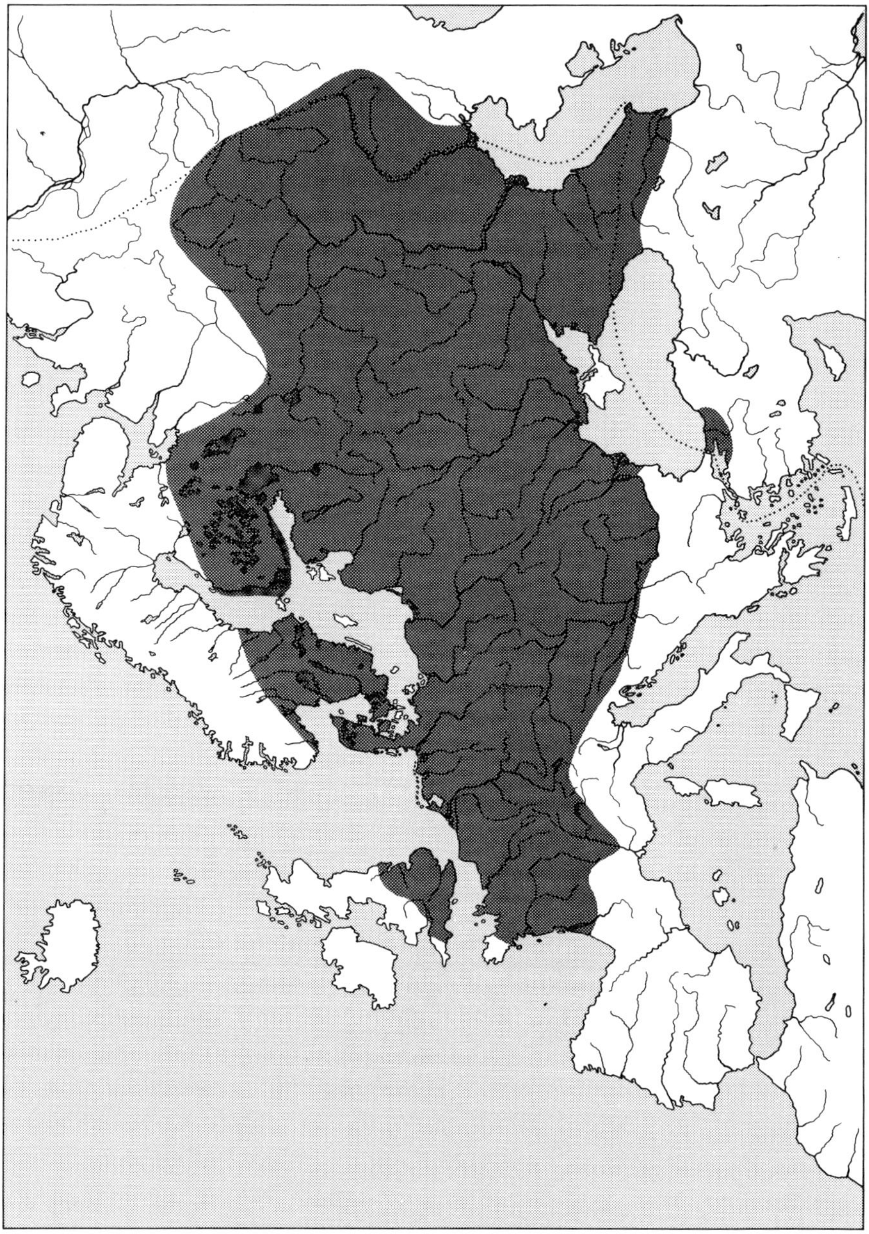

Fig. 37: European distribution of *Blicca bjoerkna* (Linnaeus, 1758).

ally in June; it is a phytophilous spawner. Unlike roach and bream, its adaptation to a variety of substrate is not particularly pronounced. The males occupy the spawning grounds first, the ripe females appear later. The eggs are released in several batches. Freshly hatched larvae provide food for earlier spawners, particularly perch, pike and roach. The males reach maturity in the second or third year of life, females a year later. Growth is slow, the maximum size reached is about 25 to 30 cm in total length. Its feeding habits are similar to those of the blue bream, but it prefers planktonic to benthic food. The mouth is not protrusible, like that of *Abramis*. The species generally belongs to the fish community of the potamal zone.

Population in Europe

Common within its distribution area, in waterbodies lacking larger predators which it tends to overpopulation, as it is not particularly valuable either for fishery or sport fishing.

Reasons for decline

There is no danger of its disappearance, though it is more sensitive to man-made environmental changes and pollution than the more common *Abramis brama*. On this reason local reductions of populations may happen.

Conservation measures taken and proposed

No conservation measures have been taken and there is no need to propose any. The re-establishment of multispecies fish populations with predators in the lower reaches of large streams would prevent local overpopulation.

Carassius carassius (Linnaeus, 1758)

E. Crucian carp
F. Caraccin
G. Karausche

Vernacular name: –

Family: Cyprinidae

Classification: rare to vulnerable

Note: A stunted form *Carassius carassius* morpha *humilis* (Heckel, 1840) is found in eastern Europe.

Distinguishing characteristics

The crucian carp is a scaly high-bodied, robust-looking fish. The mouth is small and terminal; it has no barbels. The dorsal fin is long and convex with a slightly serrated third (hard) spine. The crucial distinguishing characteristic is the number of gill rakers (22 to 33). The peritoneum is not pigmented. Meristic data: D 17 to 25, A 7 to 11; 32 to 35 (36) scales in lateral line; (6) 7 to 8 scales above the lateral line.

Distribution in Europe

There are several theories about its indigenous distribution in Central and Western Europe. Recent limits of its range are found in England and southern Scotland, crossing north-east of France. It lives in all waters connected with the Baltic and North Seas, including southern Scandinavia. Its range covers Central Europe to the Alps. In southern Europe it is found in an area in southeast of the Iberian Peninsula, but not in the Rhône and its tributaries. It is rare in the Po catchment area, and does not occur elsewhere in Italy. It is found on the Balkan Peninsula in the catchment areas of the Vardar and Maritsa Rivers. It is not found east of the catchment area of the Lena River in Siberia. See distribution map and Libosvarsky (1962).

Population in Europe

Populations of crucian carps were typical for the large flood-plains of Europe. Natural habitats were reduced by drainage of the wetlands and populations became limited to scattered pools along rivers. In northern Europe, there were stable populations in ageing lakes. Very little is known of the present distribution of populations within the natural range.

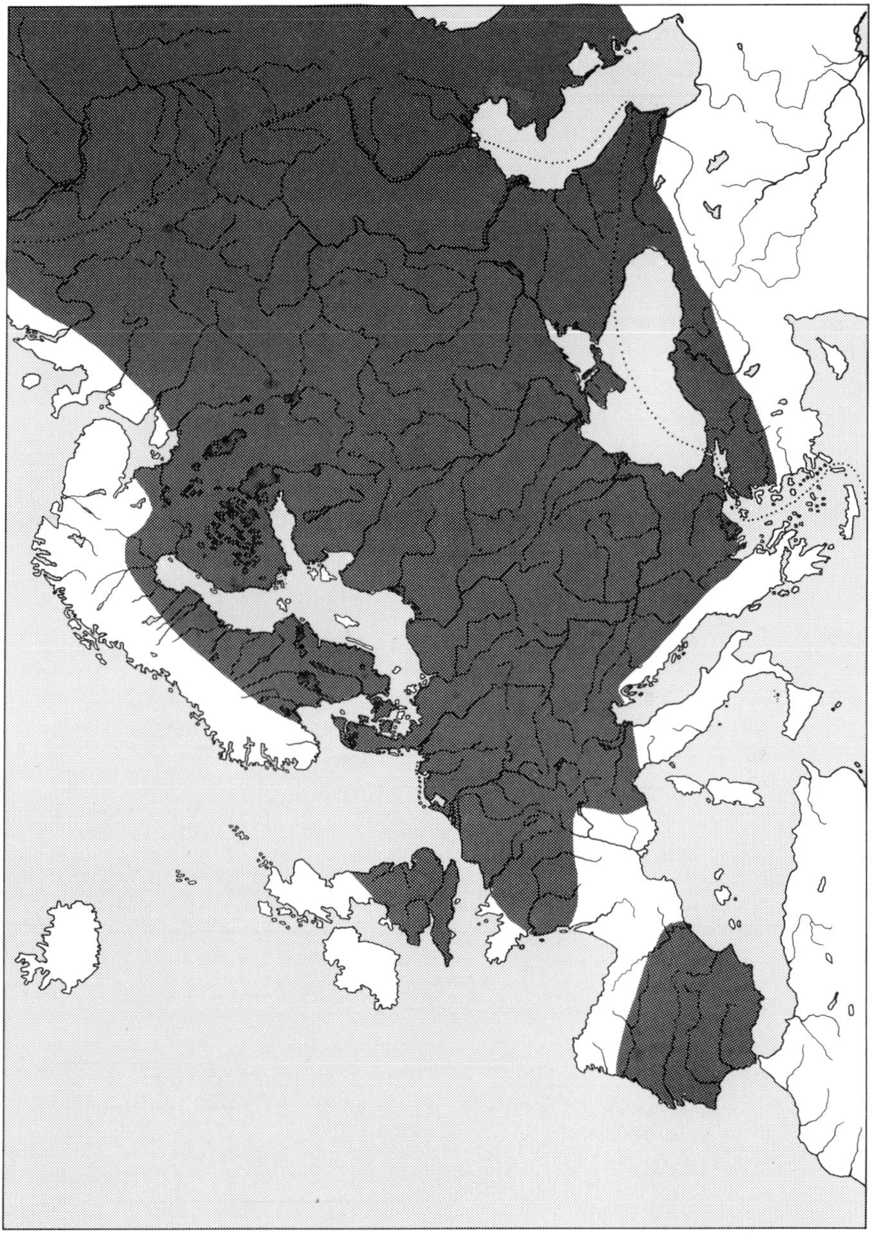

Fig. 38: Distribution of *Carassius carassius* (Linnaeus, 1758) in Europe and adjacent regions.

Ecology

This species is probably the most resistant to extreme limnological conditions. It can survive in swampy areas with a minimum of water, almost without oxygen, according to HEUSCHMANN (1957). It prefers densely vegetated pools with seasonally fluctuating water levels. It is very rare in slow rivers with muddy bottoms. Maturity is reached in the 3rd (males) or 4th (females) year of life. Spawning takes place in dense water weeds. The eggs are laid in successive batches on the vegetation. Feeding habits are very diverse, from algae to benthic fauna, but it prefers plankton. Growth rate is slow; in the Danube, 10 year-old individuals reach a length of about 20 to 25 cm. In ponds, its growth rate is faster.

Reasons for decline

The decline of this species has not been described for several reasons: Its habitats have only little importance for fishery, thus its decreasing occurrence was not noticed. Furthermore, this species very closely resembles the species *Carassius auratus* which is notoriously misidentified as *C. carassius*. The actual decline is linked with the draining of wetland pools, and, very probably *C. carassius* is gradually replaced by *C. auratus* in and along large rivers (Lusk et al. 1977). Accidental "introductions" of *C. auratus,* both as escapes from ornamental ponds or as escaped bait-fish, have probably contributed to its present wide distribution enhancing thus the replacement of the *C. carassius* by another species looking much alike.

Conservation measures taken

No conservation measures known.

Conservation measures proposed

As this species has such modest requirements regarding the physical properties of water (except pollution), conservation of existing habitats where it is still found may be sufficient for its protection.

Carassius auratus (Linnaeus, 1758)

E. Goldfish
F. Poisson rouge
G. Giebel, Goldfisch

Vernacular name: –

Family: Cyprinidae

Classification: intermediate

Distinguishing characteristics

This species looks very much like the crucian carp, but it is less hump-backed, and more elongated. Its dorsal fin is slightly concave, the peritoneum (body cavity) is pigmented, sometimes dark. The third dorsal spine is serrated, often deeply, and the scales are a bit larger. In wild forms, the colour varies. It has no dark spot in front of the tail fin. The free living varieties which derived from races, bred for ponds and aquarium, display the full range of colours, from golden-green or brown to spotted greenish-red or black. The crucial distinguishing characteristic is the number of gill rakers, which range form 39 to 50. Meristic data: D 17 to 22, A 7 to 9; 27 to 34 scales in the lateral line; 5 to 7 above it.

Worldwide distribution

There are still many uncertainties about the distribution of this species. Until recently the typical "Goldfish" *Carassius auratus,* and the "Giebel" *Carassius auratus gibelio* were separated as two taxa. It is now believed that these two forms are only geographic varieties of the same species, the former occurring in the Far East (Amur River catchment area), the latter from Siberia to northern Germany and the Black Sea. If we accept that differences between the forms are not strong enough to separate them in two subspecies (Hensel, 1971) the obvious differences in the reproduction cycle may be considered as adaptations to different environments within the large distribution area. Moreover, old records state that the Far Eastern form was brought to Europe as an aquarium fish and has been released or escaped into natural water bodies. *Carassius auratus* has also been introduced to the North American continent.

Distribution in Europe

From the Iberian Peninsula (undoubtedly artificially introduced there)

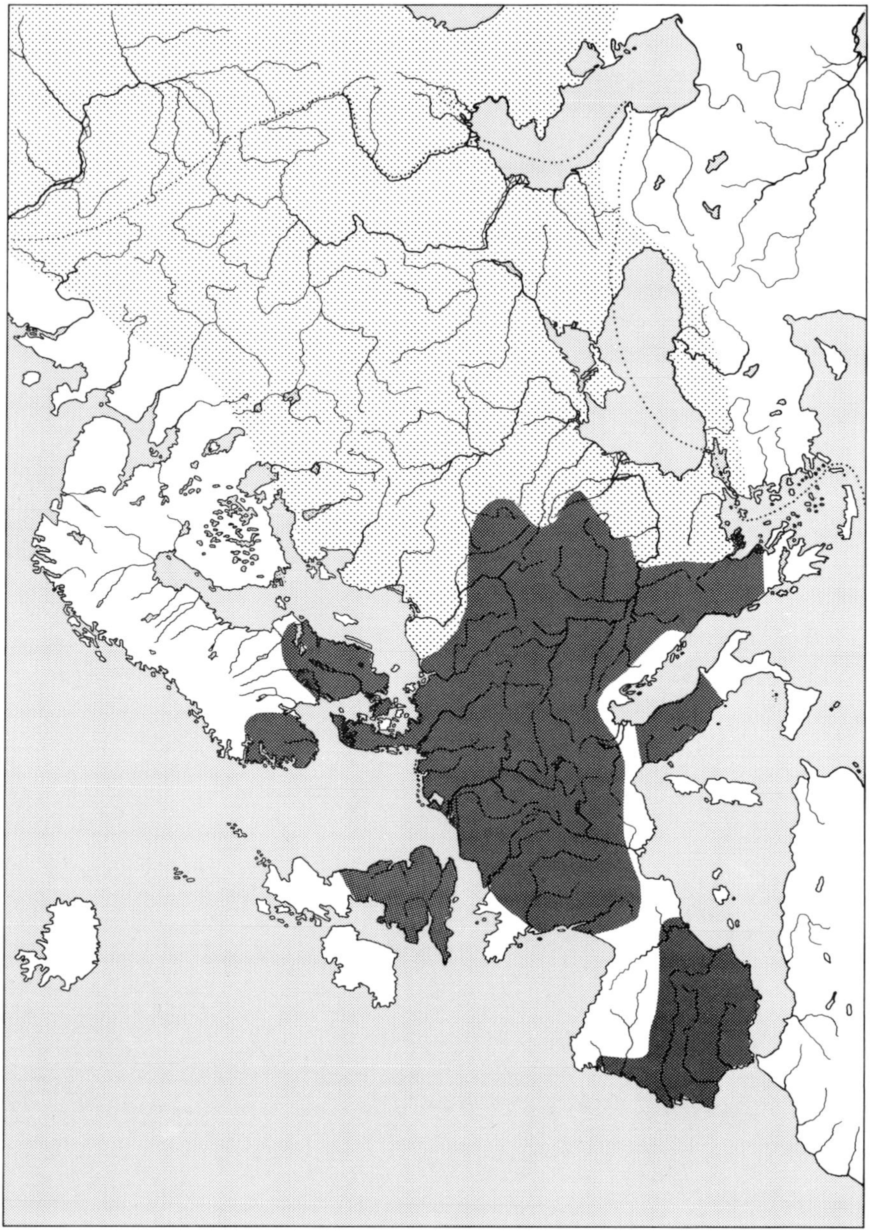

Fig. 39: European distribution of *Carassius auratus* (Linnaeus, 1758).

across Central Europe (with the exception of the northern Adriatic watershed) to the Black Sea area and eastward to Asia.

Ecology

The ecology of this species is similar to that of the crucian carp. It prefers shallow, weed grown oxbows and pools in flooded areas, shallow natural lakes with much dense shore vegetation and slow flowing streams with muddy bottoms and diversified shore line. Unlike the crucian carp, it can live permanently in streams. In general, it is not a fish of absolutely lentic river stretches, but is more rheophilous, requiring larger water bodies (crucian carp can live in small pools of several square metres!). The feeding and spawning habits of both are the same, but the growth rate of the goldfish is slightly higher, and it has a greater resistance to unfavourable limnological conditions. There is, however, one peculiarity in its reproduction: One of the two morphologically almost indistinguishable forms, the Siberian form, *gibelio,* reproduces gynogenetically, i.e. females may spawn with males of other cyprinid species but the sperm only initiates the development of eggs, without actual fertilization. The progeny is again the "giebel" but only of the female sex. Males are very seldom found among the populations; this reproduction process, described here very briefly, is known only from the variety *Carassius auratus gibelio*. The typical goldfish, *Carassius auratus,* spawns normally, as does the crucian carp. Penaz et al. (1979) reviewed the literature and contributet with experimental proofs.

Population in Europe

In spite of difficulties in determining whether or not this species is a native fish (see Libosvarsky, 1962), its distribution area is rather peculiar. It was first observed in the vicinity of cities of Central Europe to which Asian fishes have been imported. This species is now scattered throughout Europe apparently due to release or escape of captive individuals. A continuous drift of this species, comparable to population expansion, was recently noted in several Moravian tributaries to the Danube (Lusk *et al.* 1977). At the moment the range of this species in Europe is expanding.

Conservation measures taken

Not known and none required at present.

Conservation measures proposed

No protective activities are suggested, as this species might compete with native crucian carp. The reason for including this species in this volume is, that the habitats, required by this species, may undergo changes by anthropogenic influence soon. In this case populations would be influenced adversely.

Chalcalburnus chalcoides (Güldenstädt, 1772)

E. Danubian bleak

F. –

G. Seelaube, Mairenke

Vernacular name: Shemaja (Russian)

Family: Cyprinidae

Classification: vulnerable to endangered

Note: This species has several variations which are sometimes considered to be subspecies. In Central and in Southeast Europe, only *Chalcalburnus chalcoides mento* (Agassiz, 1832) is important. Other subspecies occuring in Europe are:
Chalcalburnus ch. derjugini (Berg, 1923) found in the Crimea
Chalcalburnus ch. istanbulensis (Battalgil, 1941)
Chalcalburnus ch. mandrensis (Drensky, 1943), found in Lake Mandra in Bulgaria
Chalcalburnus ch. mentoides (Kessler, 1859) in freshwaters in the Crimea;
other taxa are distributed in the Caspian and Azov Seas areas.

Distinguishing characteristics

The Danubian bleak has a streamlined body form strongly resembling that of the bleak *(Alburnus alburnus)*. The superior mouth is typical. The body is laterally compressed, the ventral part behind the pelvic fins being shaped like a scaleless keel. The dorsal part is dark-green, the sides and belly are silvery white. It may reach a length of 30 to 35 cm. Meristic data for the subspecies *Chalcalburnus ch. mento:* D 9 to 11, A 17 to 20, pharyngeal teeth in two rows; (58) 59 to 65 (68) scales in lateral line.

Distribution in Europe

The centre of its distribution area is in the rivers, emptying into the Black and Caspian Seas. The subspecies *Chalcalburnus ch. mento* (in German "Seelaube" or "Mairenke") occurs in the Danube Delta and near the Iron Gate, about 650 km upstream. Though generally considered as a Danube fish, there are no records of recent finds in the upper reaches. It is further recorded from lakes in and near to the Alps: Traunsee, Attersee, Mondsee, Grundlsee, Wolfgangsee, Würmsee, Tachensee, Simssee, Ammersee, Starnberger See and Chiemsee and in the rivulet Wörnitz in Bavaria. There are no reliable records of its recent occurrence. See map of the probable distribution of Danubian bleak and subspecies, p. 176.

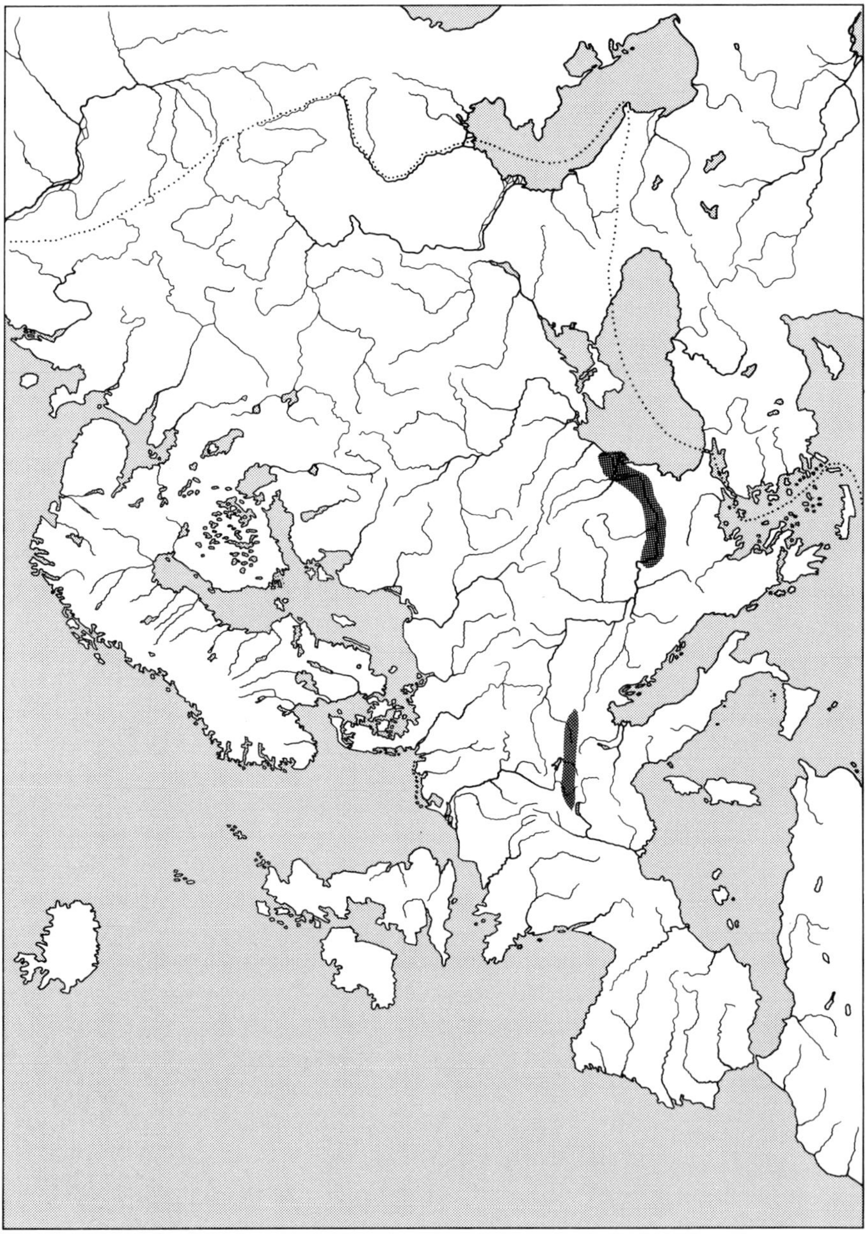

Fig. 40: Distribution centers of *Chalcalburnus chalcoides* (Güldenstädt, 1772) in Europe. For details see text, p. 175.

Ecology

In general, very little is known. Individuals living in lakes prefer cold waters above stony bottom. During the summer, they prefer the deep water. Spawning time is May and June. The fish congregate in the shallow waters of streams connected to the lake. The eggs are deposited on stones in running water. This species may reach a length of 35 cm in its 9th year of life. Its food is thought to be plankton, but it is believed that it takes drifting food as well, because its mouth position is adapted to surface feeding. The ecology of populations in the lowest Danube stretches is similar to that of the bleak.

Population in Europe

The population in Europe seems to be concentrated in several localities. The future of the Danube population after construction of the man-made lake in the Iron Gate is uncertain. There are no dependable data from the Subalpine lakes.

Reasons for decline and conservation measures

Not much information on its decline is available. If this species actually spends most of the year in deep water layers, it may become endangered by increased eutrophication. It needs attention from ichthyologists, fishery biologists, and possibly fishery mangers, who should try to obtain more information on its life cycle to make conservation measures effective.

Chondrostoma nasus (Linnaeus, 1758)

E. Nase
F. Hotu
G. Nase

Vernacular name: –

Family: Cyprinidae

Classification: intermediate to vulnerable

Note: There are at least 5 subspecies native to Europe:
Chondrostoma nasus borysthenicum (Berg, 1914) in the Dniester, Don, and Danube
Chondrostoma nasus ohridanum (Karaman, 1924) in Lake Prespa
Chondrostoma nasus vardarense (Karaman, 1928) in the catchment areas of the Wardar and Maritsa Rivers;
Chondrostoma nasus variabile (Kakowlew, 1870) in the eastern tributaries of Volga, Kama and Ural River.

Distinguishing characteristics

Chondrostoma nasus has a slim body, moderately compressed laterally. The protruding snout is fleshy, and the position of the mouth inferior. The upper lip is fleshy, while the lower one is hard and covered with a horny sheath forming a sharp chisel edge. The mouth opening is transverse, and only in small fish less than about 5 cm in length is it slightly arched. It has no conspicuous colour except during the spawning season, when the skin covering the opercular bones and the insertions of the pelvic and pectoral fins turns orange red. Meristic data: D 11 to 14, A 13 to 14; 56 to 63 scales in the lateral line; pharyngeal teeth in one row, usually six on each side.

Distribution in Europe

Distributed northward from the Alps in Central Europe; abundant in the Danube catchment area and in some rivers flowing to the North Sea, with the exception of the Elbe River system. It also occurs in the watershed of rivers north of the Baltic Sea and further eastwards to the rivers north of the Caspian Sea. See the distribution map.

Ecology

Chondrostoma nasus lives in swiftly flowing streams, in the barbel zone. Very often the stretches occupied by grayling are reached and it even as-

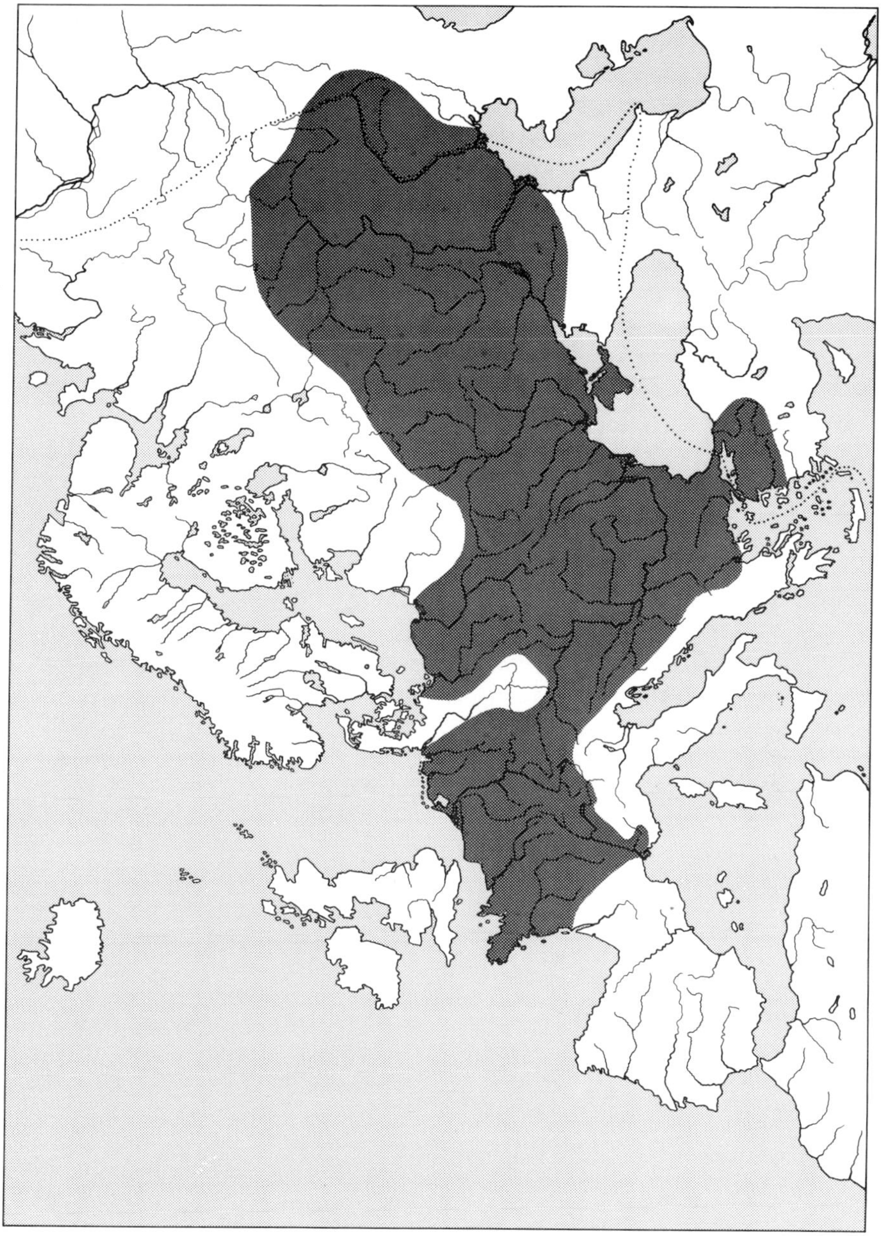

Fig. 41: Distribution of *Chondrostoma nasus* (Linnaeus, 1758).

cends to the lower reaches of the brown trout zone. It requires oxygenated water, but can withstand relatively high temperatures (over 20° C) in summer. It is a typical bottom feeder, living in schools and grazing on plants, particularly green algae and diatoms growing on stones. Spawning takes place in early spring at water temperatures of about 10° C. The fishes migrate into tributaries of the main stream and spawn in shallow water, with a maximum depth of 30 cm, on gravel substrate. Females release the eggs in one batch after very lively spawning activity with a school of males, usually more than 15 individuals. For details on spawning, see LELEK and PEŇÁZ (1963). The prerequisite for a successful life cycle and high abundance is the availability of lotic stretches with very diversified bottoms and shore line, as well as the accessibility to shallow tributaries with suitable spawning ground. Maturity is reached after the 3rd or 4th year of life, but *Chondrostoma nasus* may live more than ten years and reach a length of 35 to 40 cm and the weight of 1 to 1.5 kg. For more details see the comprehensive life history study by LUSK (1967).

Population in Europe

The nase is abundant within its range, provided the water course is not modified, regulated or polluted. Modifications, such as the construction of large dams or weirs, may cause its almost gradual disappearance from waters where it was formerly abundant, as in the lower stretches of the Upper Rhine. It is seldom harvested commercially. For unknown reasons, its introduction into not native but suitable habitats is difficult, as was observed in some upper tributaries of the Elbe in Central Europe.

Reasons for decline

Changes in the river flow, particularly caused by canalization and dams, and the extension of lentic river stretches, has eliminated this species from parts of its distribution area.

Conservation measures taken and proposed

This species is now protected during its spawning season almost everywhere. It is recommended that fish sanctuaries be established to guarantee its survival (see comments on the barbel p. 159 and section 4, p. 316).

Chondrostoma toxostoma Vallot, 1836

E. Soiffe
F. Seuffe, Toxostome
G. Südwest-europäischer Näsling

Vernacular name: Svetta (Swiss)

Family: Cyprinidae

Classification: vulnerable

Note: One subspecies, *Chondrostoma t. arrigonis* (Steindachner, 1866) occurs in the Jucar Basin (Spain) and Southern Portugal (in the Tejo, Guadina Rivers), (see Collares-Pereira, 1978)

Distinguishing characteristics

This species shows all the characteristics, typical for the genus *Chondrostoma*. It has a dark stripe on the side of the body, stretching above the lateral line from the end of the opercular bone to the insertion of the anal rays. This stripe may not always be clearly visible. The Svetta may reach a length of 25 cm. Meristic data: D 11, A 12; 54 to 59 scales in lateral line. (This species may be conspecific with *Chondrostoma genei*. Further investigation of systematics and taxonomy seems urgent).

Distribution in Europe

The distribution area stretches from the northeastern slopes of the Pyrenees, and includes the Ebro river basin and the Rhône in the south. North of the Pyrenees, it occurs in the Tejo, and perhaps in the Loire in France. See distribution map, p. 182.

Ecology

This species lives in small rivers, and occasionally in lakes with clean, well-oxygenated water. It is probably permanently present in upper river sections where *Chondrostoma nasus* only spawns. Spawning takes place in early spring, when the water temperature reaches 13° C, always in gravelly shallows upstream from its normal habitat.

Conservation measures: See *Chondrostoma genei*.

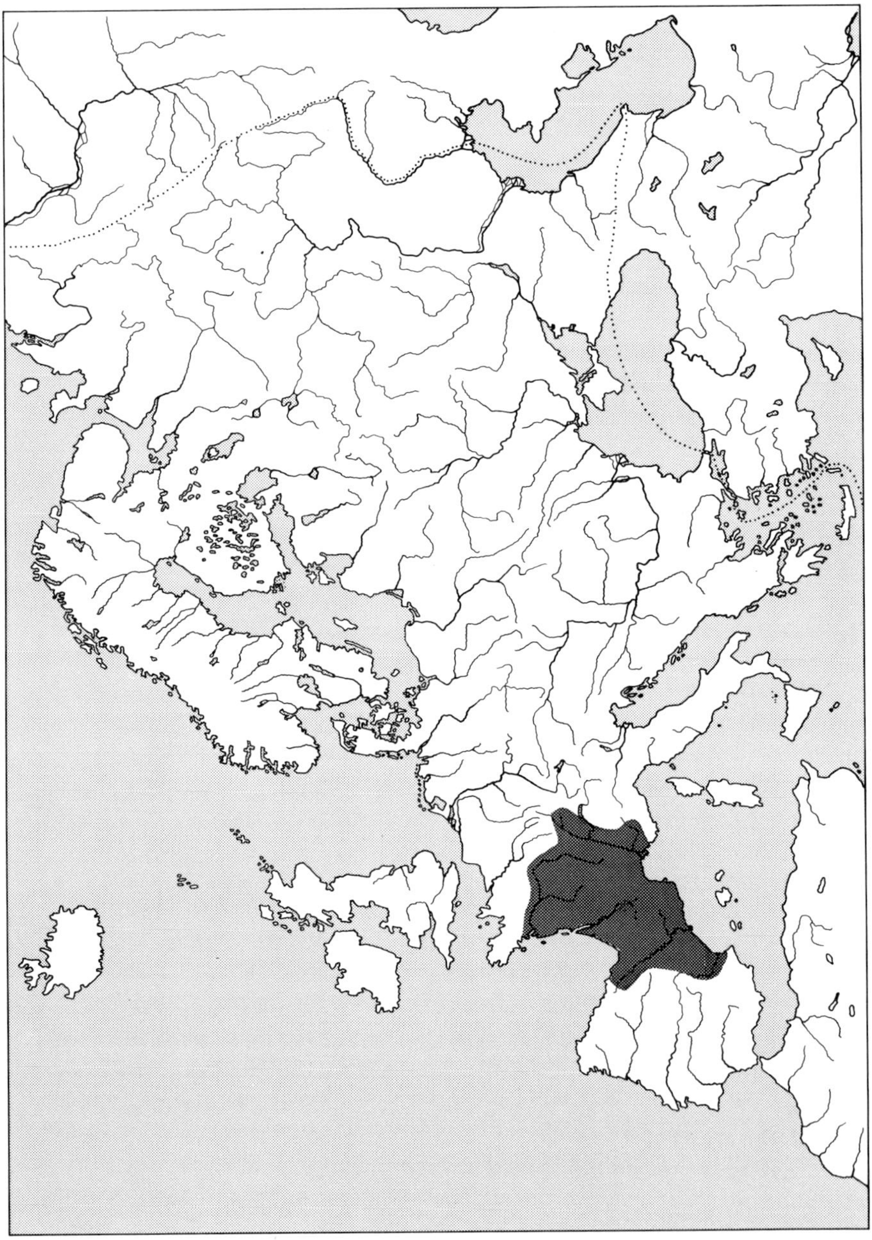

Fig. 42: Distribution area of *Chondrostoma toxostoma* Vallot, 1836.

Chondrostoma genei Bonaparte, 1832–1841

E. –
F. –
G. Lau

Vernacular name: Casca (Italian)

Family: Cyprinidae

Classification: vulnerable

Note: The validity of the subspecies *Chondrostoma g. albicans* (Gatti, 1896) seems dubious. No comparative material is available for examination of the status of this fish, which occurs in the rivers emptying in the Adriatic Sea.

Distinguishing characteristics

The general body habitus is the same as that of *Chondrostoma nasus,* including the mouth position. There is a conspicuous dark band on the sides, extending from the pectoral fin, located above the lateral line, to the insertion of the caudal fin. Meristic data: D 11, A 11–12, pharyngeal teeth in one row, 5 to (exceptionally) 6 teeth. It is believed that this fish is conspecific with *Chondrostoma toxostoma.* Further investigations are urgently needed.

Distribution in Europe

Ch. genei occurs from north to central Italy, in the Adige and Po catchment areas, and supposedly in the Varo River. Its occurrence in the Upper Rhine, Inn and Danube catchment area is extremely uncertain, as no old and recent records are available. A tentative distribution map appears on page 184.

Ecology

The habits of this species are thought to be similar to those of *Chondrostoma nasus,* though it is less dependent on algae.

Conservation measures

No conservation measures are proposed because there is very little information about this fish available. It is strongly urged that the taxonomy, distribution and ecology of this species be thoroughly investigated.

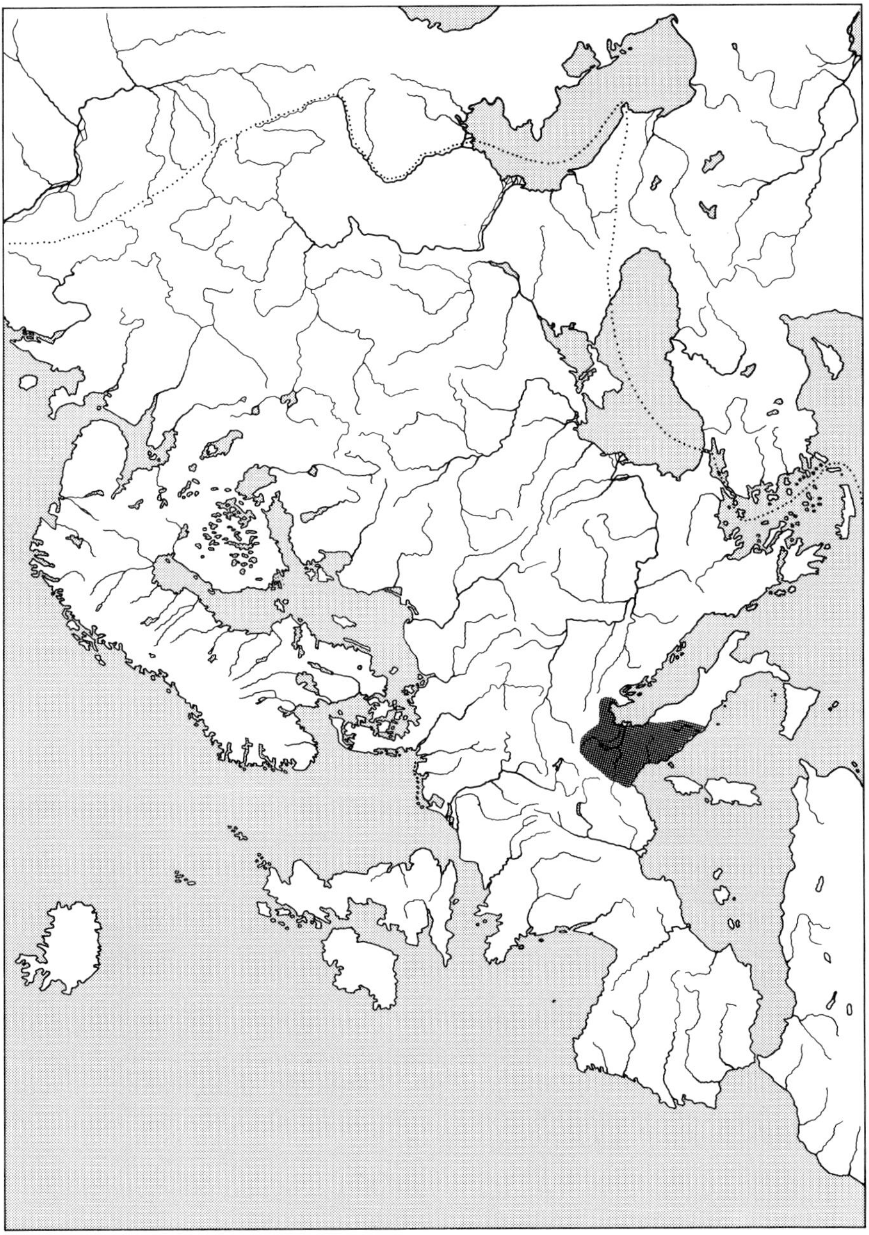

Fig. 43: Distribution area of *Chondrostoma genei* Bonaparte, 1832–1841.

Chondrostoma soetta Bonaparte, 1832–1841

E. –
F. –
G. Italienischer Näsling

Vernacular name: Savetta (Italian)
Family: Cyprinidae

Classification: vulnerable

Distinguishing characteristics

This taxon has all the features of the genus *Chondrostoma,* but it is more corpulent than other species of this genus. The dorsum is blue-green, turning to black on the sides, and the belly white, often with a yellow tinge. The fins are dark. It may reach a length of 45 cm. Meristic data: D 12, A 14 to 15, pharyngeal teeth in two rows, 7 (6) in each row.

Distribution in Europe

It occurs in rivers south of the Alps, including the Po, Adige, Brenta and Piave Rivers. See distribution map, p. 186.

Ecology

This species forms schools and is a bottom feeder, taking both animal and plant food. Otherwise, its ecology is comparable to *Chondrostoma nasus.*

Population in Europe

No exact data are available.

Conservation measures

No conservation measures are taken or proposed. Nothing is known on changes of populations. It is probably endangered on ground of its small range.

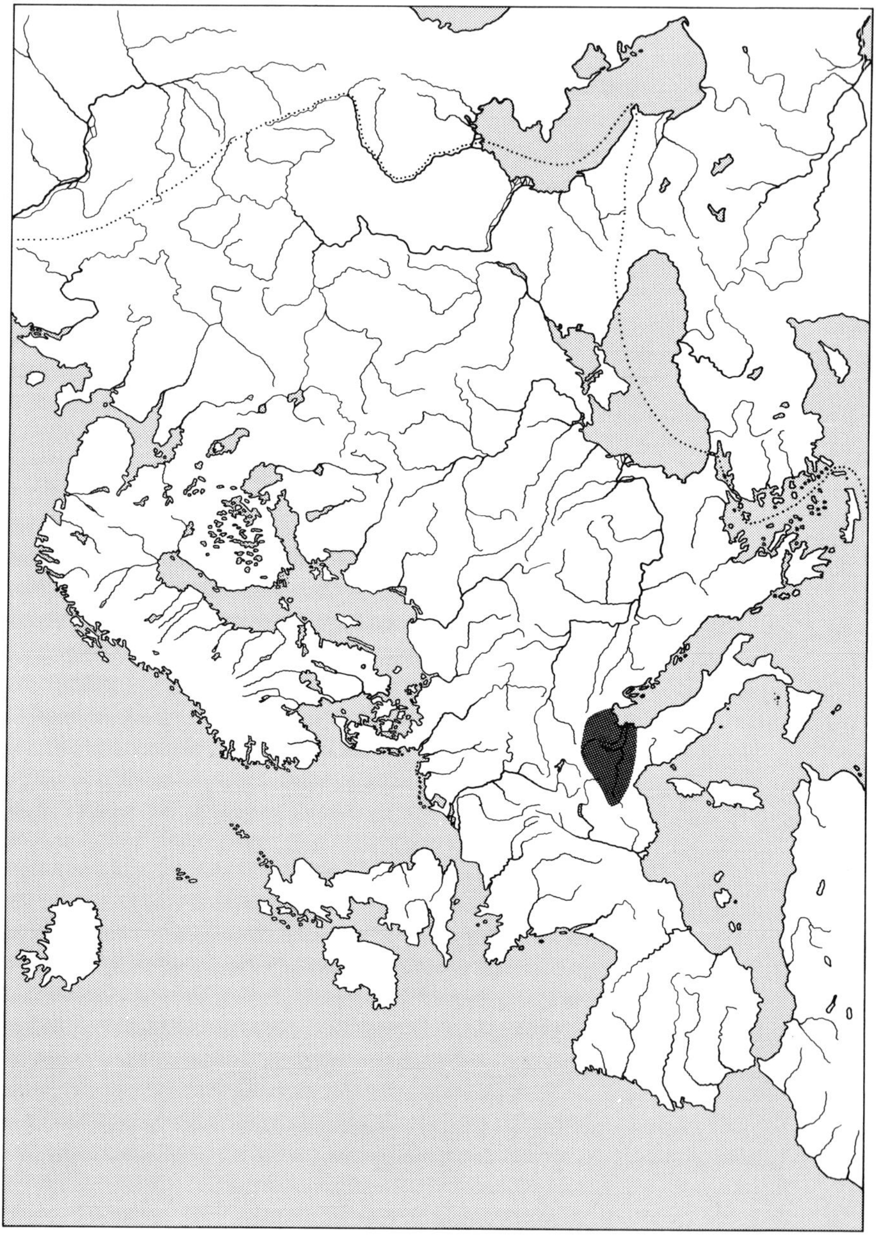

Fig. 44: Distribution area of *Chondrostoma soetta* Bonaparte, 1832–1841.

Gobio albipinnatus Lukasch, 1933

E. –

F. –

G. Weißflossiger Gründling

Vernacular name: Beloperyi peskar (Russian)

Family: Cyprinidae

Classification: rare

Note: This species is represented in Europe by 3 subspecies. The typical form, *Gobio a. albipinnatus* lives in the Volga and Don catchment areas. *Gobio a. belingi* (Slastenenko, 1934) occurs in the rivers Dnjester, Bug and Dnieper. *Gobio a. elimeius* Steph. occurs in the catchment area of the Pimos River, Greece. *Gobio a. vladykovi* Fang, 1943 is treated in more detail, as this form occurs in the European area considered here.

Distinguishing characteristics

This species resembles the gudgeon (*Gobio gobio* p. 190). It has a scaleless throat, its barbel reaches the hind edge of the eye and the snout is short. The dorsal and caudal fins are not irregularly marked, the black dots usually being arranged in one clear transverse stripe on the caudal fin. As a rule, the second stripe ist not as clear as the most posterior one. The lateral line is marked with two parallel lines, formed by narrow dark spots. On the sides 8 to 10 dark blotches, which sometimes appear bluish, can be seen. The margins of scales on the dorsal part of the body have an epithelial keel. The dorsal fin is very pale with only a few irregular dark spots. It is a small fish, reaching a maximum length of about 12 cm. Meristic data: D (8) 9 to 10, A (7) 8 to 9; 40 to 45 scales in the lateral line and 4 (5) scales below.

Distribution in Europe

This species is found only in East-Central and Eastern Europe. Its westernmost location is the Danube in Austria and Germany. The taxonomic identity of these populations was reviewed by Balon *et al.* (1987). Downstream it is found in the Danube to the Delta in many tributaries. See distribution map.

Ecology

The ecology of *Gobio albipinnatus* is basically similar to that of the gudgeon, *G. gobio*. The differences are as follows; it is dominant in main

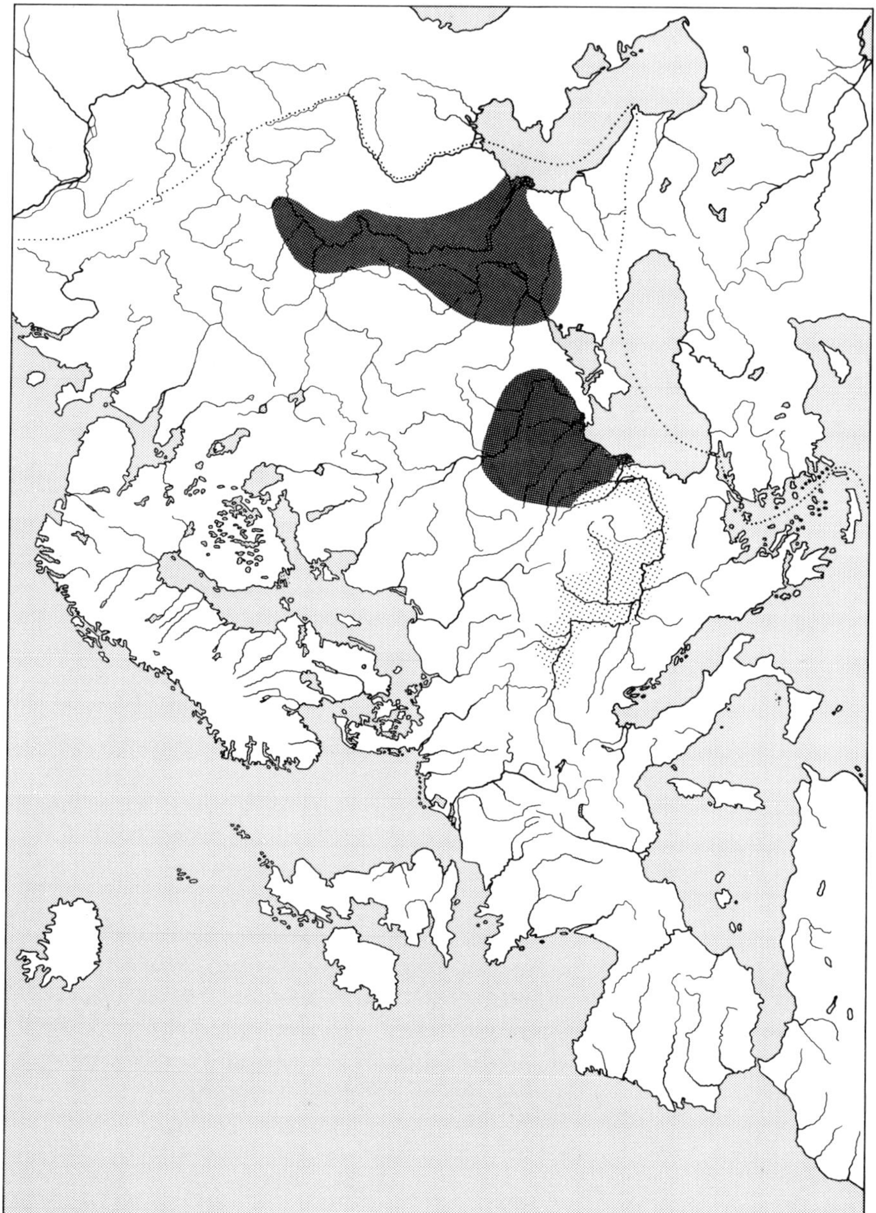

Fig. 45: Distribution area of *Gobio albipinnatus* Lukasch, 1933, (dark area) and the subspecies *Gobio a. vladykovi* Fang, 1943 (dotted area)

stream of rivers, but does not reach the upper stretches. It occupies deeper, slowly flowing waters with soft bottoms. It often occurs in oxbows adjacent to the main stream, and in some Danube tributaries from Czechoslovakia (e.g. River March), it is more abundant than the gudgeon itself.

Population in Europe

As it is a small fish, unattractive both economically and for sport fishing, very little information is available.

Reasons for decline

The same as for the gudgeon (see p. 193), but it is hoped that this species will be less sensitive to silting of the river bed after damming of water course. The predation pressure of the eel may be primarily responsible for its decline in the upper Danube.

Conservation measures taken or proposed

No conservation measures known. In general, the same protection as required for the gudgeon.

Gobio gobio (Linnaeus, 1758)

E. Gudgeon
F. Goujon
G. Gründling

Vernacular name: –

Family: Cyprinidae

Classification: intermediate to rare

Note: The distribution area of this species is extremely large, as is the variety of habitats in which it may occur. This explains the many environmental adaptations. Many authors have tried to separate this variable species into subspecies. Some of the opinions of Bănărescu (1964) and Ladiges (1978) are partly accepted here, bearing in mind that only European taxa are considered here:

Gobio g. obtusirostris (Cuvier and Valenciennes, 1844) occurs in the Danube, though between the upper and central Danube stretches

Gobio g. gobio is also present. In the lower stretches *Gobio g. bulgaricus* replaces *G. g. obtusirostris*.

Gobio g. bulgaricus (Drensky, 1927) occurs in the lower Danube, and the Maritsa and Wardar basins.

Gobio g. albanicus (Oliva, 1960) is found in the Balkan peninsula.

Gobio g. faraeensis Steph. occurs in Grecian waters.

Gobio g. kovatschevi (Chichkoff, 1937): eastern part of Bulgaria, Danube catchment area.

Gobio g. ohridanus (Karaman, 1924) occurs in Lake Ohrid.

Gobio g. krymensis (Deljamure, 1937): in the Crimea.

Gobio g. holurus (Berg, 1914): in eastern Transcaucasia.

Gobio g. caucasicus (Kamensky, 1901) in the Rion river basin. 8 species and subspecies from Asian Turkey, some of which may be dubious forms (see Ladiges, 1960), are not mentioned here.

Distinguishing characteristics

The body has the general shape of a bottom-dwelling fish, with more or less straight ventral side and a slightly convex dorsum. In cross-section, it is round anteriorly and compressed laterally toward the tail. The mouth is inferior, with one barbel in each mouth corner. Both anal and dorsal fins are short-based. It reaches a maximum length of 16 to 20 cm. The body colour is very variable and depends on the environment. The dorsum is always dark, ranging from brownish-grey to brownish-green. The colour becomes lighter on the sides and yellowish-white on the belly. On the sides there is a row of large, indistinct, more or less rounded dark marks. The dorsal and the tail fins are irregularly spotted; the dots on the tail fin may be arranged in five regular vertical rows. In some exceptional specimens the barbels

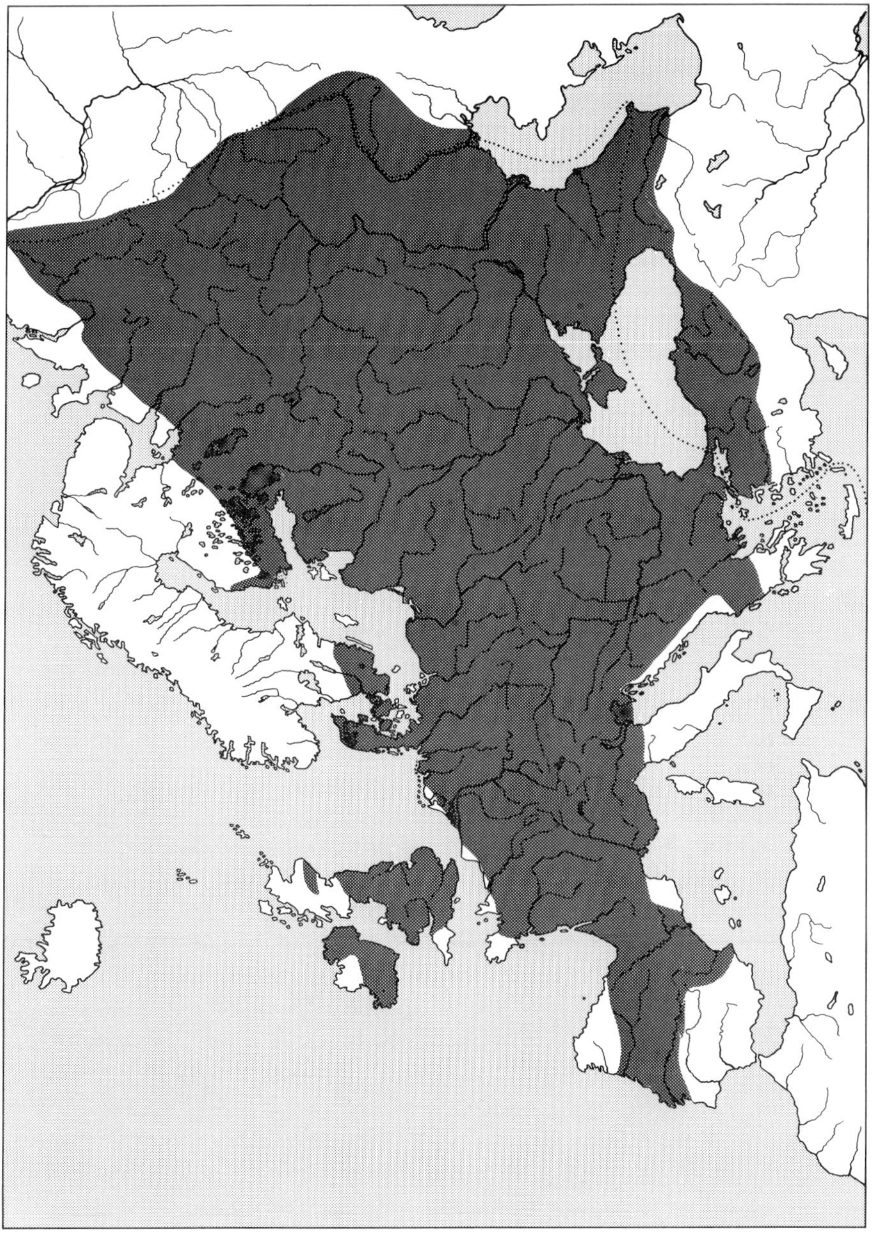

Fig. 46: General distribution of *Gobio gobio* (Linnaeus, 1758) including several subspecies.

may reach the hind edge of the eye. The throat is scaleless. Meristic data: D 10 to 11, A 8 to 10; 40 to 45 scales in the lateral line, 5 to 6 scales above it; pharyngeal teeth in two rows.

Distribution in Europe

The gudgeon occurs in great parts of Europe. In Ireland, England and Scotland as far north as the river Dee. It is absent from Norway, northern Sweden and Finland. The original southern range limit was north of the Pyrenees. It was introduced into Spain, and the range is now expanding over the Douro and Tejo river systems to Portugal (REY 1947, ALMAÇA 1965). It occurs throughout Europe, across the Black Sea basin to the rivers feeding the Caspian Sea from the North and West. In the South, it occurs only within the northern Adriatic watershed, with the exception of Dalmatia, Greece and southern Turkey.

Ecology

The gudgeon is a non-migratory species with moderate oxygen and water quality demands. In spite of its modest environmental requirements, this small species can indicate the deterioration of the water body with a high degree of sensitivity. Essential for this species is suitable spawning substrate, in swiftly to moderately running water with a clean bottom. It does not matter whether this is sand or small gravel, but it must not be covered with silt. This species seems to be a typical psammophilous spawner. The larvae are phototrophic. This enables them to avoid crevices or sheltered bottom niches where oxygen depletion may occur sooner than elsewhere. The gudgeon occurs in small creeks, sometimes even within the distribution area of the brown trout. It is most abundant in the middle sections of streams. However, a lentic environment does not deter this species, provided the spawning substrate is available. Sexual maturity is reached in the 2nd (males) or 3rd (females) year of life; maximum life expectancy is 5 to 6 years. Adults can live on soft bottoms, often covered with silt and mud. Food is mainly benthic animals. In lentic environments (ponds and lakes) young individuals take plankton. *Gobio gobio* can withstand rather high water temperatures in summer, often above 25° C, provided the water remains oxygenated.

Population in Europe

Before rivers were changed into uniform canals, the gudgeon was one of the most abundant species in European rivers. At present, its occurrence depends on suitable spawning grounds, from which it spreads also into less suitable habitats as juvenile fish. In many large streams, where it was formerly common, it has disappeared almost completely from long stretches, e.g. the Main and Rhine Rivers.

Reasons for decline

Man-made modifications to streams, pollution of any kind other than municipal sewage and loss of suitable spawning places are responsible for drastic local decline of this species.

Conservation measures taken

No conservative measures taken so far.

Conservation measures proposed

This species may be protected and conserved permanently only by protection of the environment. As *Gobio gobio* does not undertake real spawning migrations, but moves only over short distances, protection of small localities with suitable spawning substrate may be sufficient to save the remaining isolated populations.

Gobio kessleri kessleri Dybowski, 1862

E. Kessler's gudgeon

F. –

G. Kesslerscher Gründling

Vernacular name: Porcusor demisp (Romanian)

Family: Cyprinidae

Classification: rare and vulnerable

Note: Two other subspecies are recognized: *Gobio k. antipai* Bănărescu, 1953, occuring in the Danube Delta and its lower branches and *Gobio k. banaticus* Bănărescu, 1953 which inhabits the Danube tributaries in Hungary, Yugoslavia, and Romania.

Distinguishing characteristics

This species has the basic gudgeon body habitus and is well adapted to bottom dwelling. The pigmentation of the tail fin is remarkable: there are two rows of barely visible dark dots running parallel with the edge of the fork. The posterior row is always conspicuously darker than the anterior one. The barbels reach the hind edge of the eye. The caudal penduncle of the tail is always rounded; the throat is scaleless. The colouration is conspicuous. It is brighter than that of *Gobio gobio,* turning to silver. Below the lateral line there are only 3 (exceptionally 4) scales. Meristic data: D (9) 10 to 11 (12), A 8 to 9 (10); 40 to 42 scales in the lateral line.

Distribution in Europe

Only the distribution of *Gobio k. kessleri* is considered here. Other subspecies occur in the Danube Delta and the Dniester basin. The westernmost distribution seems to be the March River in Moravia and the Danube in the vicinity of Bratislava (Slovakia). Further eastward, it is found in Danube tributaries. The upper tributaries of the Vistula form the northern range limit. It has not been recorded in the upper Danube during this century (Terofal, 1977). At present, in the subspecies *antipai* and *banaticus* was observed extension of their distribution (pers. comm. P. Bănărescu and L. Michajlowa).

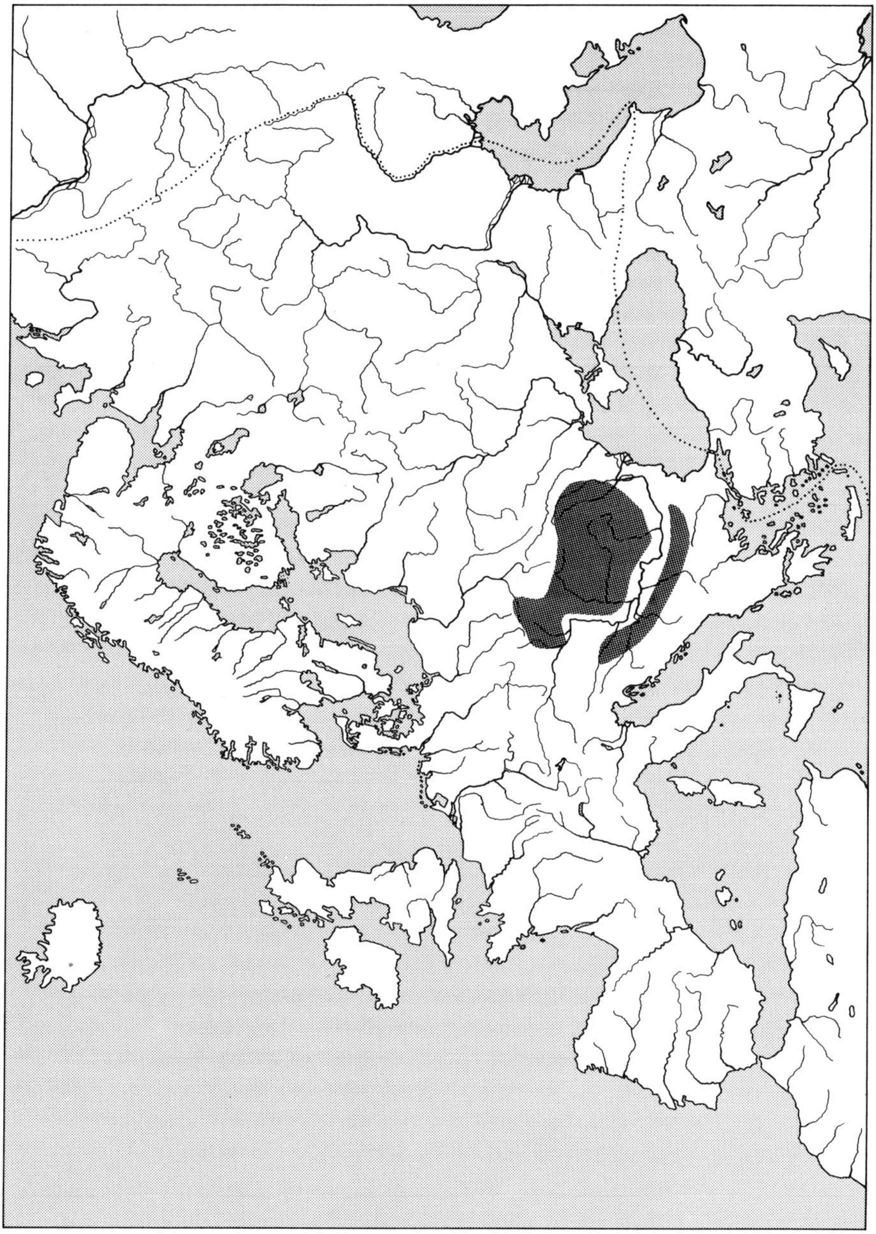

Fig. 47: Distribution of *Gobio kessleri kessleri*, Dybowski, 1826.

Ecology

Much less is known on the life of *Gobio kessleri* than about other species of this genus. It is rather solitary, living in swiftly flowing stretches with stony and gravelly bottom upstream from the habitat of *Gobio gobio*. It is a typical bottom-dwelling fish.

Populations in Europe

This species has probably always been rare.

Reasons for decline and conservation measures

Because of its scarcity, any alteration of the physical environment may be detrimental to this fish. To guarantee its future existence, at least some localities where it is still found should be protected form anthropogenic changes.

Gobio uranoscopus (Agassiz, 1928)

E. Danubian gudgeon
F. –
G. Steingressling

Vernacular name: –

Family: Cyprinidae

Classification: rare and vulnerable

Note: One subspecies is known, *Gobio u. frici* (Vladykov, 1925), described from the easternmost part of Czechoslovakia (now USSR). Its validity, however, seems to be dubious. The different mean number of scales in the lateral line is an insufficient reason for classifying it as a subspecies.

Distinguishing characteristics

This fish has the basic "gudgeon habitus", but is the slimmest member of the genus. The caudal region of the body, and particularly the caudal penduncle, is rounded. The barbels are very long, reaching beyond the preopercular bone. The throat is scaly. The pigmentation of the caudal fin is similar to that of *Gobio kessleri,* but is often more conspicuous. The dark stripes in the caudal fin are clearly visible. It has 4 to 5 dark, relatively large, blotches on the dorsum which extend ventrally almost to the lateral line. The body colour is dark, the sides yellowish-grey. Meristic data: D 9 to 10, A 7 to 9; 40 to 43 scales in the lateral line.

Distribution in Europe

This species is confined to the Danube Basin. The northern distribution limit is the Isar and Salzach Rivers, and it can be expected in the upper reaches of the Danube itself, though this has not yet been confirmed. Further downstream it has been observed in Slovakia, Yugoslavia, Hungary, and Romania. It inhabits the north-eastern tributaries of the Tisza River in the USSR. See tentative distribution map based mostly on BĂNĂRESCU (1975).

Ecology

The Danubian gudgeon lives only in the rapids of the grayling and barbel zone, where currents range from 0,5 to 1,5 m/sec. The water quality must not drop below the oligosaprobic level and it must be well saturated with

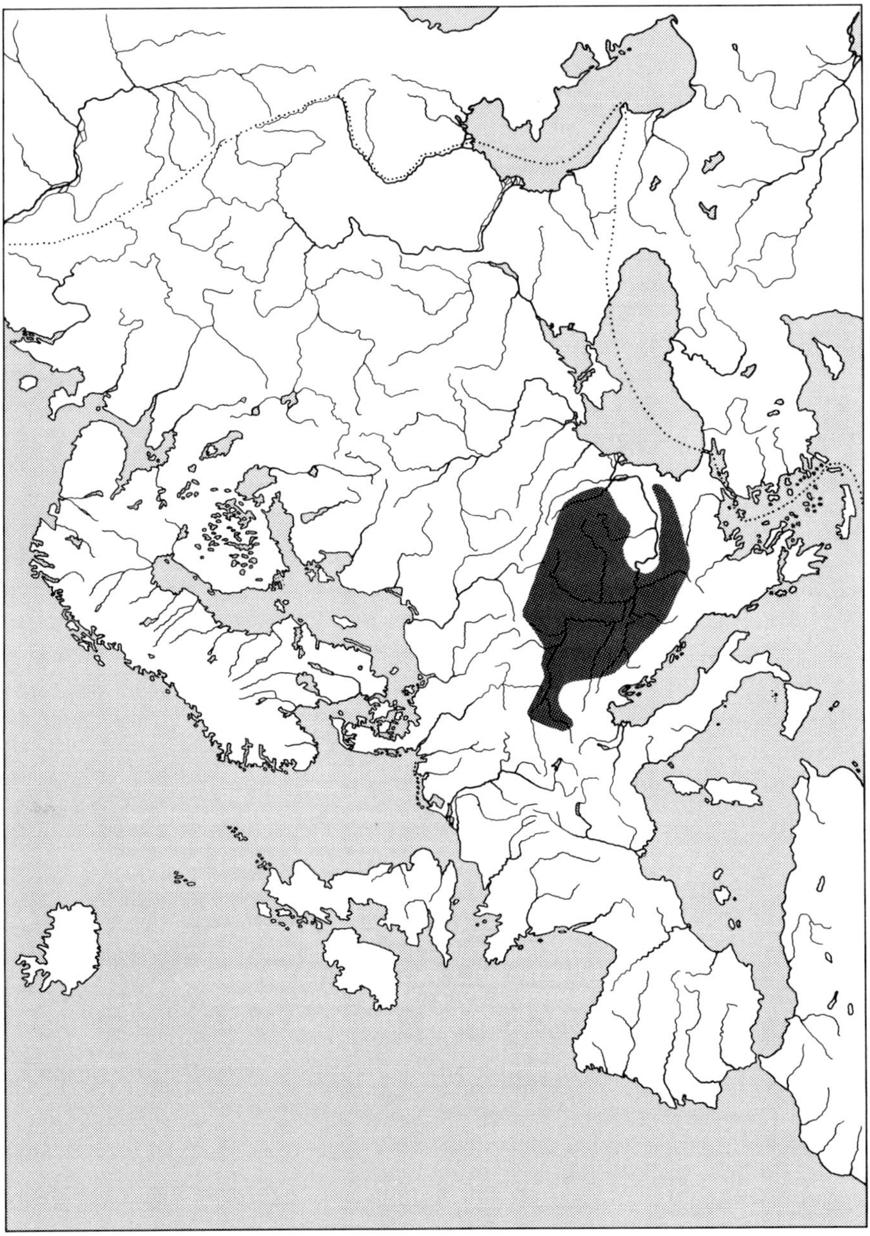

Fig. 48: Distribution of *Gobio uranoscopus* (Agassiz, 1928).

oxygen. Other requirements are similar to those of *Gobio kessleri*. This species seems to be the most sensitive of all "gudgeons". No further information on its ecology is available.

Population in Europe

It is clear that this species is nowhere abundant, and the population strength is not known.

Reasons for decline and conservation measures

As it is not known how abundant this species was in the past, the present scarcity should not be considered as indication of a recent decline. If this fish in fact requires conservation, its biotopes must be preserved. Any changes in water discharge or environment and pollution load must be prevented.

Leucaspius delineatus (Heckel, 1843)

E. Sunbleak (Moderlieschen)
F. Able de Heckel
G. Moderlieschen

Vernacular name: –

Family: Cyprinidae

Classification: rare to vulnerable

Note: A related form from western Turkey is sometimes considered a species or occasionally as a subspecies: *Leucaspius irideus* (Ladiges, 1960). It lives in western Anatolia: other subspecies (or species) may exist in this area, but are not confirmed.

Distinguishing characteristics

The sunbleak is a small-bodied fish with large, thin, fragile scales. The lateral line pores are found only on the first 2 to 12 scales. The mouth is superior, and the lower jaw projects beyond the upper one. The average length of adult individuals is 5 to 7 cm. The body is brilliant silver, the dorsum is olive, with a silvery blue band running along the sides from head to tail. Meristic data: D (10) 11 to 12, A 13 to 16.

Distribution in Europe

General distribution in Europe begins at the Rhine River and continues eastwards. Its artificial distribution probably reaches the English Channel. It occurs in some left-hand tributaries of the Rhine, some lentic habitats along the Main River and in canals along the Saone River. It is not found in the Alps, the Balkan Peninsula, the British Isles or Denmark, but it is present in southern Sweden and throughout Europe to the east. Its range ends in the catchment of the Dvina and Volga Rivers. It is not found in the Crimea, nor in the Maritsa or Vardar River Basins. Related forms occur in western Anatolia and in Greece. See distribution map, p. 201.

Ecology

Leucaspius delineatus is a gregarious fish, living in lentic or very slowly flowing waters with vegetation on the shores. Very often, it occurs in irrigation channels if there is little or no current. It prefers waters overgrown with vegetation, but with some open surface in the middle. *Leucaspius* be-

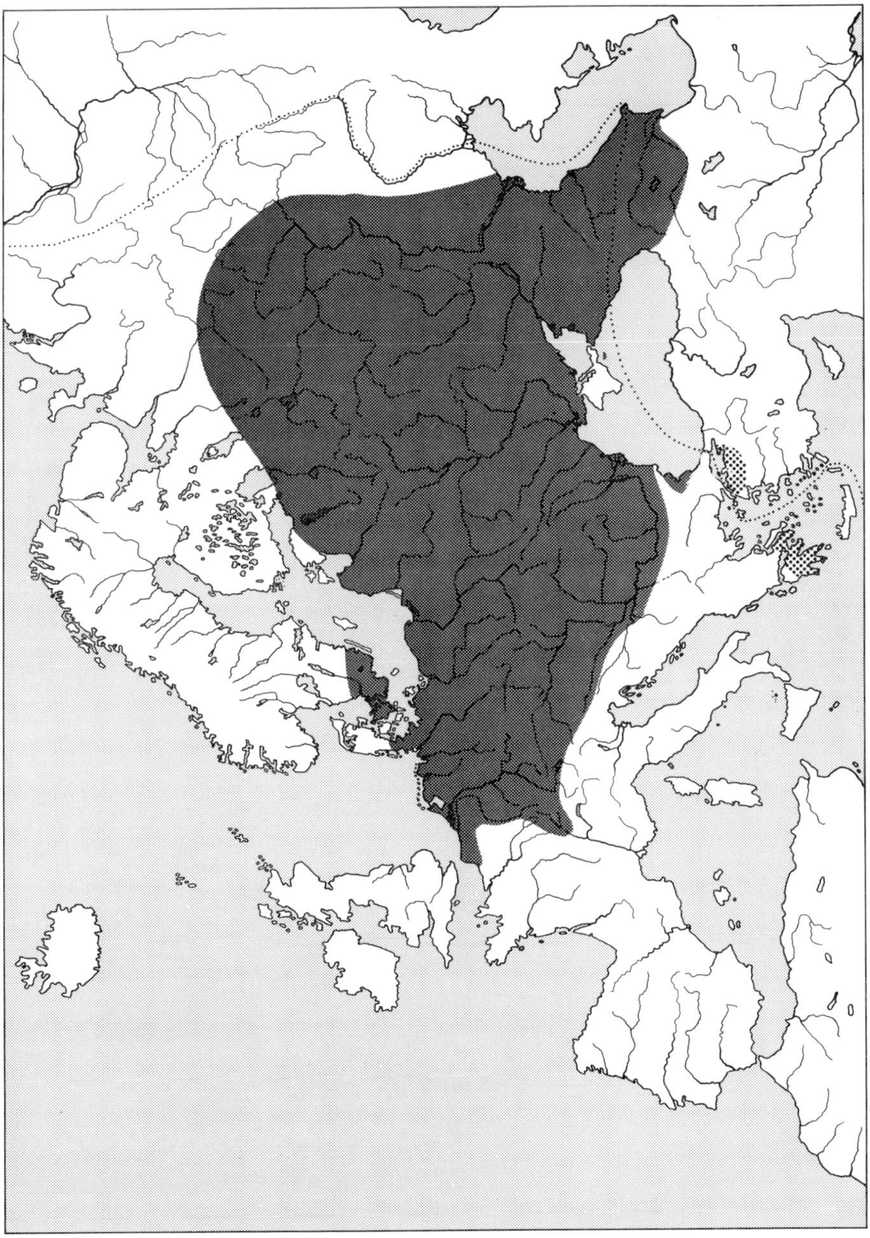

Fig. 49: Distribution of *Leucaspius delineatus* (Heckel, 1843).

longs to the phytophilous, guarding fishes. The female attaches ribbon-like threads of eggs to the lower side of broad leaves, around stems or floating objects. The clutch is protected by a primitive nest structure made of vegetation, and guarded and aerated by the male during the incubation period. The average life span is short, 2 to 3 years, and the maximum life expectancy is about 4 or 5 years. The largest individuals reach a length of 7 to 8 cm. The fish is sensitive to pollution but not to fluctuations in temperature and water level.

Population in Europe

As this species has specific environmental demands, the populations are scattered in the channels and the pools near to larger water bodies and in canals connecting ponds in which carps are raised.

Reasons for decline

The drainage of wetlands for cultivation has contributed to the disappearance of many channels in areas subject to flooding. Improvements, and particularly draining have exterminated many populations in Central Europe. Large scale application of insecticides and herbicides may also have caused local disappearance of this species.

Conservation measures taken and proposed

No conservation measures are known and protection of this species can be achieved only by means of biotope conservation.

Leuciscus idus (Linnaeus, 1758)

E. Orfe, Ide
F. Ide mélanote
G. Aland, Nerfling

Vernacular name: –

Family: Cyprinidae

Classification: vulnerable to endangered

Note: The ide has probably one subspecies named *Leuciscus idus stagnalis* (Dubalen, 1913) found in southern France in the vicinity of the Rhône estuary, but the validity of this taxon is dubious. A red variety kept in garden ponds and also used as an indicator organism in pollution control is well known: *Leuciscus idus* aberratio *orfus*.

Distinguishing characteristics

The ide has the usual *Leuciscus* body habitus, but looks more corpulent, and the smaller head makes it appear a little humped. The snout is blunt; the mouth oblique and terminally set. The colour is conspicuous. The dorsal part of the body is dark, turning to silver on the sides and white on the belly, contrasting with bright red fins, the anal being the brightest. Meristic data: D (10) to 11 (12), A 12 to 13 (15); 55 to 63 scales in lateral line; 8 to 9 rows above it.

Distribution in Europe

The ide is common in Central and East Europe, reaching the Lena River Basin in the East. It was introduced to England and Scotland, where it occurs sporadically. It is found in Scandinavia, but not in northwest Norway or the Kola Peninsula. In Central Europe it occurs from Belgium and northeastern France (Escaut River) to rivers feeding the Black and Caspian Seas. See distribution map, p. 204.

Ecology

Leuciscus idus requires large river or lake systems in which migrations are possible: in spring, upstream to spawn in gravelly shallows; in summer, into flooded areas to feed, and in autumn downstream to deeper rivers or lakes for the winter. Maturity is reached sometime in the 2nd year of life; 3 to 5 years old individuals are mature. During its migrations from March to early

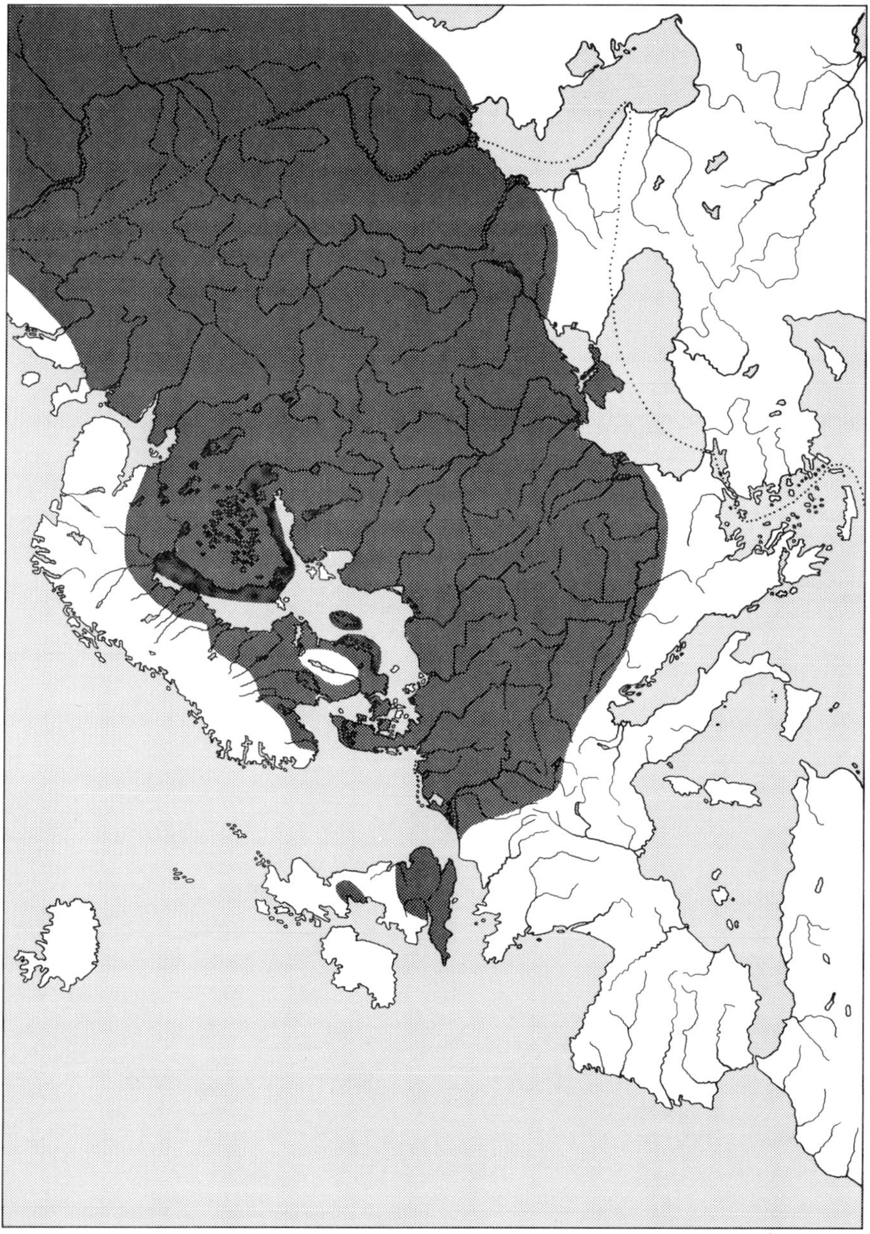

Fig. 50: European distribution of *Leuciscus idus* (Linnaeus, 1758).

May, it may be considered a gregarious fish, but during the feeding period, the adults are more or less dispersed in the feeding grounds. It often takes drifting food from the surface, but mainly benthic insects, sometimes plants and occasionally small fish. Growth is relatively fast: In the first year 6 cm, 18 cm in the third and about 30 cm in the sixth year. It may reach a length of over 50 to 60 cm, weigh more than 1.5 kg and live well over 15 years. This species may live permanently in larger lakes and reproduce there successfully, e.g. in the Masurian Lakes in Poland or small high-altitude lakes.

Population in Europe

In Central Europe stable river populations exist only in the Danube river, from southern Germany downstream. However, in this area, fluctuations in its occurrence were observed. A gradual decline in migrating individuals, as in the March and Thaya Rivers, have been observed since the Second World War. A similar situation is found in the Rhine system and the Main River, where changes in historical times, based on old records, have been summarised by KLAUSEWITZ (1974). Rather unstable populations exist in the lowest reaches of the Rhine, but solitary species are observed from the central part of this river (Bless, pers. comm.). One individual was recorded from the Upper Rhine in 1978 (LELEK, 1979).

Reasons for decline

Pollution seems to have played the most important role in the local extinction of this species, while the building of locks and weirs made the indispensable spring migrations impossible.

Conservation measures taken

In many European countries a size limit and closed seasons have been introduced to protect this species, but no ecological measures have been taken on a sufficiently large scale to be effective.

Conservation measures proposed

Firstly, the pollution load must be decreased. It is believed that a further small reduction of pollution in the Rhine would help the recovery of this species considerably. It is hoped that the rigorous conservation of areas subject to flooding and new measures to connect old river branches with the main stream will have a decisive effect. Such measures might create new centres for the recovery of endangered and extinct populations. Reintroduction of this species in formerly natural habitats is strongly recommended.

Leuciscus leuciscus (Linnaeus, 1758)

E. Dace — Vernacular name: –
F. Vandoise
G. Hasel — Family: Cyprinidae

Classification: intermeditate and often vulnerable

Note: This species occurs within Europe almost entirely as one race, though the subspecies *Leuciscus l. hurdigalensis* (Valenciennes, 1844) and *Leuciscus l. bearnensis* (Blanchard) have been described from the Garonne and southern France.

Distinguishing characteristics

The dace is a slim bodied scaly fish with a relatively small head, and a terminal or slightly inferior, narrow mouth. The dorsal fin usually has 7 branched rays; the anal fin is concave. It has no particular colour pattern. The dorsal part of the body is dark, the sides silver, the belly white. The colour of the fins is characteristic. They are never red as in *Leuciscus idus;* those on the belly are creamy yellow, though the anal fin ranges from yellow to orange. It seldom reaches more than 25 to 30 cm in length. Meristic data: D 10 (11), A 10 to 11 (12); 45 to 55 scales in the lateral line; 7 to 9 rows of scales above it. Pharyngeal teeth in two rows, 5 plus 2 on each side, occasionally 5 plus 3. They are smooth and laterally compressed.

Distribution in Europe

Leuciscus leuciscus is native to great parts of Europe. It occurs in England and part of Scotland, was introduced into a part of Ireland, but only in rivers running to the Irish Sea. In Scandinavia, it is not found in most parts of Norway nor the Kola Peninsula, but in Sweden and Finland. Further eastward it is absent only from the northernmost territories. Distribution in Southwest of Europe begins from the Pyrenees, includes France and Central Europe, including the Rhône basin, but does not include Italy and Dalmatia. Present in the whole Danube catchment area, including rivers feeding the Black and Caspian Sea from the north, but not in the lowest reaches. The dace does not occur on the Balkan peninsula and in Greece. See distribution map, p. 208.

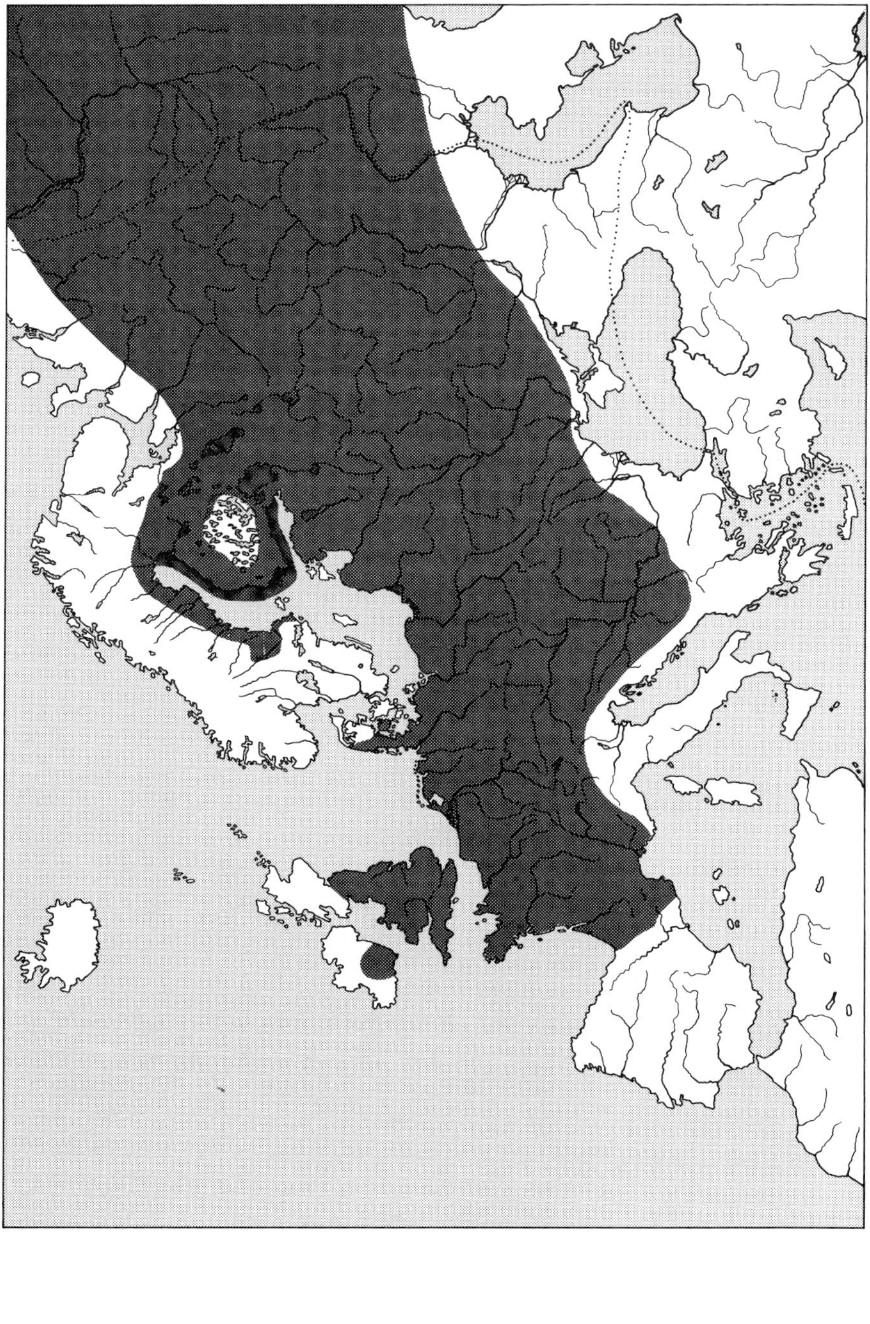

Fig. 51: Distribution of *Leuciscus leuciscus* (Linnaeus, 1758) in Europe.

Ecology

The dace is basically a rheophilous fish, but it may occasionally occur in lakes and lowland rivers. It is gregarious and spawns very early. Spawning time is connected with the geographic region: in southern England in February, in Central Europe from end of March to April, when water temperatures reach 9 to 10° C. Spawning takes place in shallows places, preferably such with gravelly substrate. Where the environment is uniform, the dace can use very small rapids (sometimes only several metres long) such as fish ladders or small torrents emptying into the main stream. It may be considered a lithophilic spawner. *Leuciscus leuciscus* is more sensitive to lack of oxygen than to pollution. On this reason it often can be found in quite heavily polluted streams as long as oxygen content is quite high. Maturity is reached early, usually in the 2nd or 3rd year. It is considered an omnivorous fish, preferring the larvae of water insects but taking drift food, as well.

Population in Europe

Where suitable spawning places are readily available, the populations are stable and the dace is abundant. In rivers with limited and scattered spawning places such as the central section of the Rhine, great fluctuation in its abundance was observed. This, combined with unfavourable weather during spawning time, results in irregular recruitment and unstable populations.

Reasons for decline

Man-made changes to the river bed and shore line, leading to uniformity in the environment and consequent loss of spawning grounds, are responsible for local reduction in population sizes.

Conservation measures taken and proposed

No conservation measures are known. On the contrary, in some areas, particularly in the trout stretches of some streams, the dace is controlled by electrofishing to prevent competition with more "valuable" fishes. In endangered river sections, only a small improvement in water quality and the protection of spawning grounds will guarantee the conservation of threatened populations. This species, within its distribution area, is in no danger, but populations in some river stretches are threatened on ground of the above mentioned reasons.

Leuciscus (Telestes) souffia Risso, 1826

E. Soufie
F. Blageon
G. Strömer

Vernacular name: –

Family: Cyprinidae

Classification: vulnerable to endangered

Note: At least three subspecies have been described. The best known are probably *Leuciscus souffia agassizi* (Couvier and Valenciennes, 1844), which will be treated in more detail here, together with *Leuciscus s. souffia*. *Leuciscus s. muticellus* (Bonaparte, 1837–38), found from north to Central Italy in the Isonzo, Po, Arno and Tiber Basins. *Leuciscus kaedicus* Steph. occurs on the Balkan Peninsula and in Greece.

Distinguishing characteristics

This species is a small, slim-bodied fish, somewhat resembling the dace *L. leuciscus*. It has a relatively small head, a slightly subterminal mouth, and thin lips. The colouration is probably the most conspicuous of all *Leuciscus* species: the sides and belly are silvery, the dorsal part dark blue. Slightly above the lateral line, there is a dark band which begins behind the eye and ends at the caudal pendulum. The bases of the paired fins have an orange tinge. During spawning, the dark stripe turns bluish-red and the lateral line itself is clearly visible, coloured almost yellow. Meristic data: D (9) 10 (12), A 11 to (13).

Distribution in Europe

The original distribution of the soufie was not quite the same as it is now. It now occurs only in some scattered areas in the Rhône Basin. The closely related subspecies *Leuciscus s. agassizi* was formerly found in the Rhine River, where it probably survived until now (Parey, pers. comm.), but there is no proof of its occurance in the Main River in recent decades. Its easternmost, isolated population seems to be that in tributaries of the Tisza River in Romania. It supposedly occurs in some Subalpine lakes. No recent reliable information on its distribution is available. See distribution map for supposed range, p. 211.

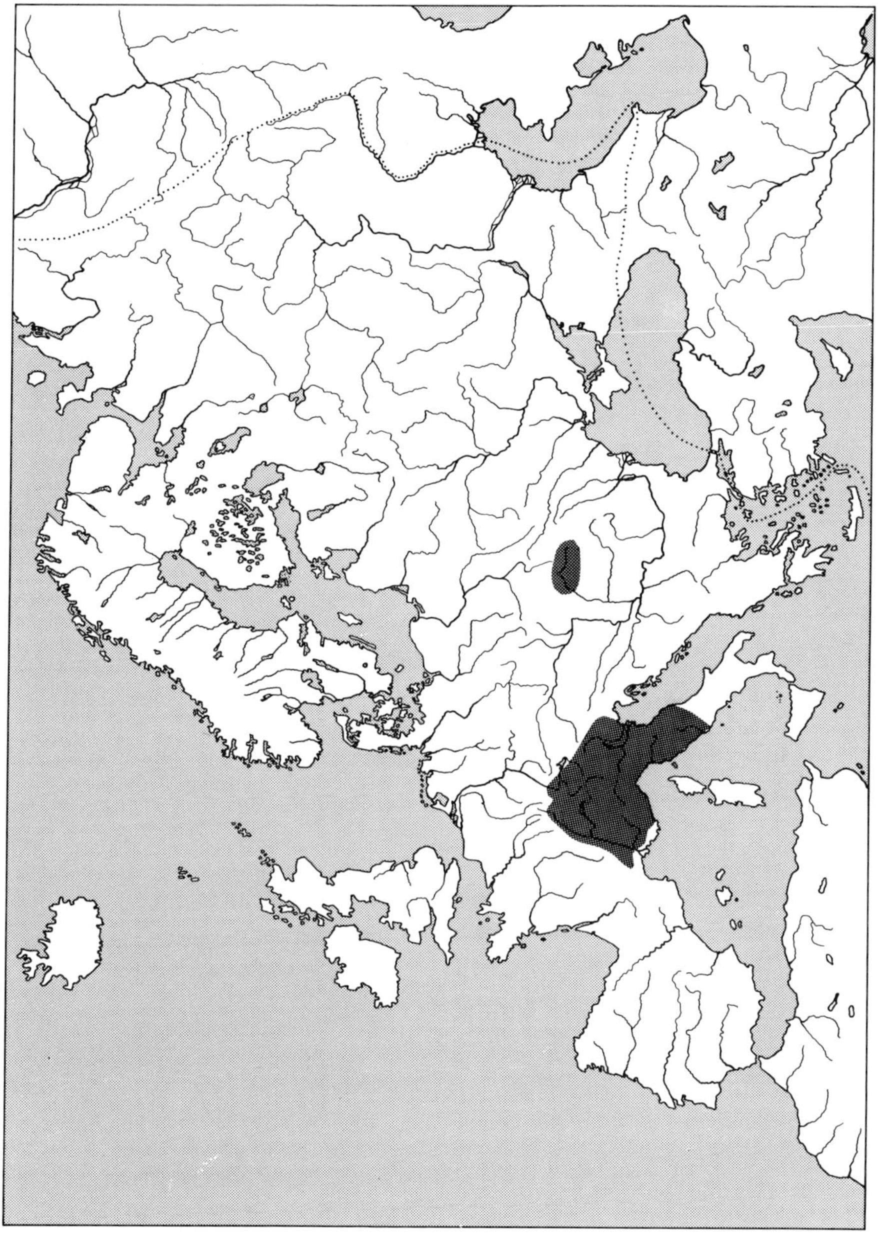

Fig. 52: Probable distribution of *Leuciscus souffia* Risso, 1826.

Population in Europe

There are no dependable data on the present distribution of this species, particularly none of *L. s. agassizi.* The sizes of the populations in the Rhône Basin are not known in detail, and no data are available concerning its distribution in other areas.

Ecology

Leuciscus souffia lives (or lived), in swift-flowing upper reaches of smaller streams, often together with salmonids. It is an early spawner and like *Chondrostoma nasus* it requires the same spawning substrate: clean, stony or sandy stretches of montane water courses, always with well-oxygenated water. It was observed to be a gregarious fish in river stretches with sandy bottoms. For information on its life cycle, see HEUSCHMANN (1957), pp. 98–101, though this information may no longer be up to date.

Reasons for decline

Only one fact is clear: this species was much more abundant in the last century than it is today. Because there are virtually no recent data on its population density, its environmental preferences or its life-cycle, it is open to speculation whether it was reduced by natural population fluctuation or because of man-made environmental changes or caused by the management of brown trout in the streams.

Conservation measures taken and proposed

This species must be fully protected with its environment. Biological data must be collected so that actual protection and conservation can be undertaken on a sound basis. The reintroduction of this species in rivers formerly inhabited would be desirable.

Pelecus cultratus (Linnaeus, 1758)

E. Sabre carp? Ziege

F. –

G. Ziege

Vernacular name: Stärkniv (Swedish)

Family: Cyprinidae

Classification: intermediate to rare

Distinguishing characteristics

This fish species is unmistakable. The body is extremely compressed laterally; the clearly visible lateral line is irregular and sinuous. The belly has a sharp keel. Other features visible at first glance are the superior mouth, the sickle-blade shape of the pectoral fins, the long anal fin, and the very short dorsal fin, situated posteriorly. Colouration: the dorsum is olive green, the sides silvery. Meristic data: D 8 to 10, A 25 to 31; 95 to 115 scales in the lateral line.

Distribution in Europe

It has two clearly separated distribution areas: one around the Baltic Sea, and the other in the area of the Black, Azov, Caspian and Aral Seas. In both regions it enters the estuaries. In Scandinavia it is found in southern Sweden and southeastern Finland. In the Danube, it occurs as far upstream as Austria and 1986 observed in the Upper Danube (Harsanyi, pers. comm.). See distribution map, p. 214.

Ecology

Pelecus cultratus is a gregarious fish, which congregates in large schools in brackish waters and lower reaches of rivers. It spawns in freshwater and is a pelagic and/or open-water-spawner. The eggs are released in the current of large rivers and the larvae hatch in 2 to 3 days. There are populations that live permanently in closed water, while spawning migrations have also been observed in the Black and Caspian Seas. Maturity is reached in the 3rd year. As a juvenile, it feeds on macroplankton, but adults prey on small fish. The average length is about 30 cm, but individuals have been known to reach 60 cm and 11 years of age.

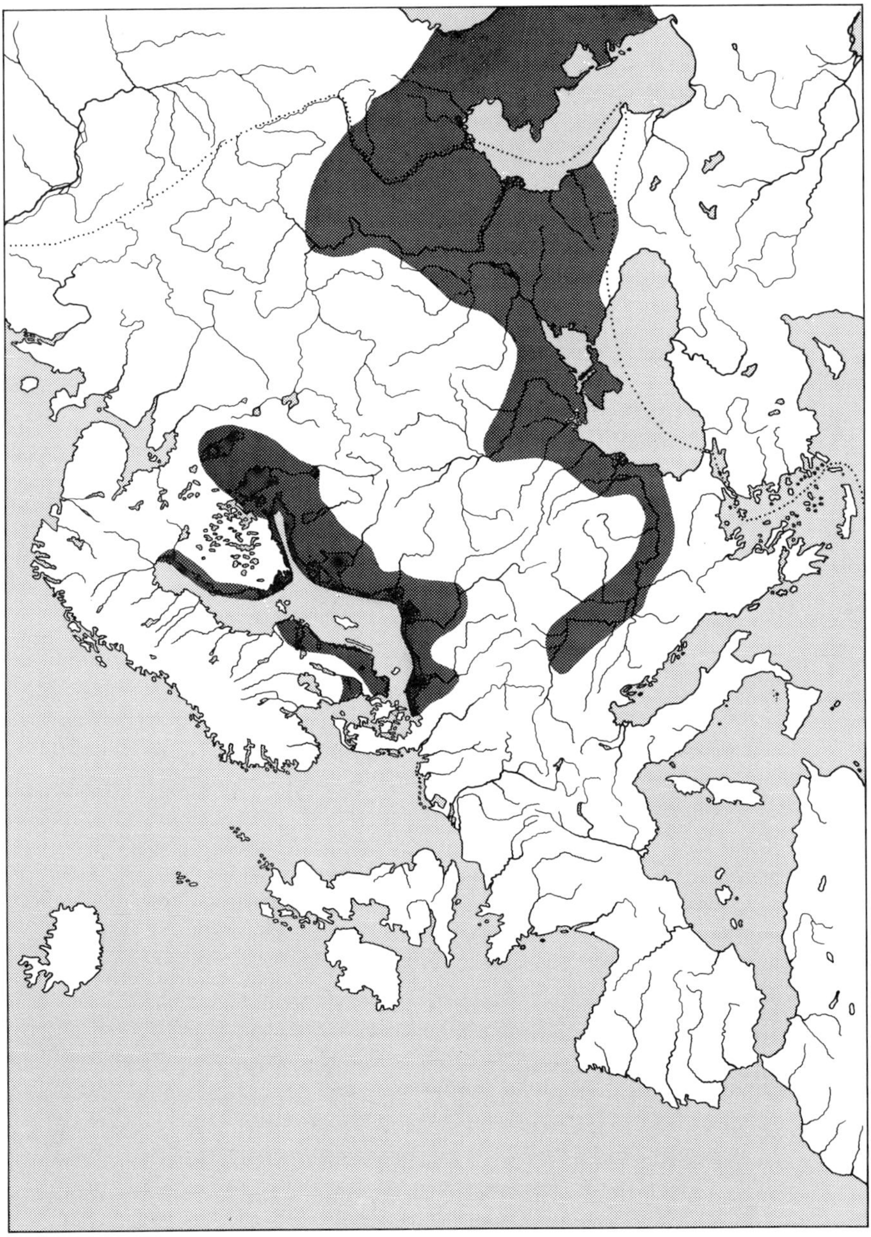

Fig. 53: Distribution of *Pelecus cultratus* (Linnaeus, 1758) in Europe and adjacent regions.

Population in Europe

In northern Europe it seems to be declining and is considered rare in the Baltic and its watersheds. It is still fairly common in tributaries of the Danube and in rivers feeding the Black and Caspian Seas.

Reasons for decline

Unknown, as it seems unlikely that pollution is responsible for its decline (HOLČÍK and HENSEL, 1972).

Conservation measures taken

Not known.

Conservation measures proposed

No fishery concentrates on this fish, but it is caught with other species. Thus, less intensive fishing would not help much. No conservation measures are proposed as little is known of its life-cycle and the reasons for decline. It is possible that the construction of power plants with dams and locks has influenced the populations and will endanger the survival of this fish in future. The population of the Volga River, however, did not suffer from the damming of the watercourse. More research on its biology is strongly recommended.

Genus *Phoxinellus*

This genus includes fishes formerly assigned to *Paraphoxinus*. The systematics and taxonomy of this group require additional research. In this group species-specific research is complicated by a high amount of synonyms. Different generic, specific and subspecific names for the same fish have been used by various authors. The recent changes in nomenclature are clearly shown by comparison of that used by LADIGES and VOGT (1965) and later on by LADIGES (1978). The name of genus *Paraphoxinus* is not used any more, and all forms are included in the genus *Phoxinellus*. It is hoped, that a taxonomic revision of this genus will take into account in addition to morphologic features the biology of species. As most forms under consideration occur in the karst of the Balkan Peninsula, the importance of investigations in this area in respect to zoogeography should be stressed. STANKOVIC (1960) seems to be the best authority for guidance on decisions on conservation measures. The *Phoxinellus* species, eg. *Phoxinellus adspersus, Phoxinellus stymphalicus* as well as *Phoxinellus hispanicus* may be renamed or regrouped in the years to come. Their presence, however, stresses the richness of forms and the geographical and zoological values of streams south of the Alps and the Pyrenees. At present, it is not clear whether these species with several subspecies are endangered in their natural surroundings. It is strongly recommended to examine the ecology and life cycles of these existent species because in case of environmental deterioration some conservation measures might be necessary.

Phoxinus phoxinus (Linnaeus, 1758)

E. Minnow
F. Vairon
G. Elritze

Vernacular name: –

Family: Cyprinidae

Classification: vulnerable

Note: Apart from the nominate form *Phoxinus p. phoxinus* one subspecies is known from rivers flowing to the Black Sea from the east: *Phoxinus ph. colchicus* (Berg, 1910).

Distinguishing characteristics

The minnow is a small fish. Its body is almost circular in cross-section, and is covered with minute scales. The lateral line is not complete in the rear section. The mouth is slightly subterminal. Colouring varies only slightly with locality. The dorsum is usually olive brown with a rich golden sheen laterally, transitioning to yellow on the belly. Sexual dimorphism is very clearly visible during spawning time: the male has a red belly, the colours of the body are darker. Tubercles are clearly visible on the top of the head and the front of the pectoral fins. The females develop only minute tubercles. Meristic data: D 9 to 10 (11), A 9 to 10; 77 to 92 scales in lateral line, pharyngeal teeth in two rows, each tooth ends in a little hook.

Distribution in Europe

The minnow has one of the largest distributions in Europe. It occurs from north of the Pyrenees to the Urals and on to the Amur Basin. It is absent from northern Scotland, western Norway, Italy (except in the north), the southeastern Balkan Peninsula, and the Peloponese. It is found in the mountains up to 2000 m altitude, occasionally even higher.

Ecology

The minnow is found in small creeks as well as medium sized streams, and occasionally in lakes with sandy bottoms. It lives in both mountain and in lowland waters, provided the water is not polluted and the bottom is scattered sandy or stony. It occurs in the river stretches together with trout and grayling, as well as in tributaries to the main streams in their lowland

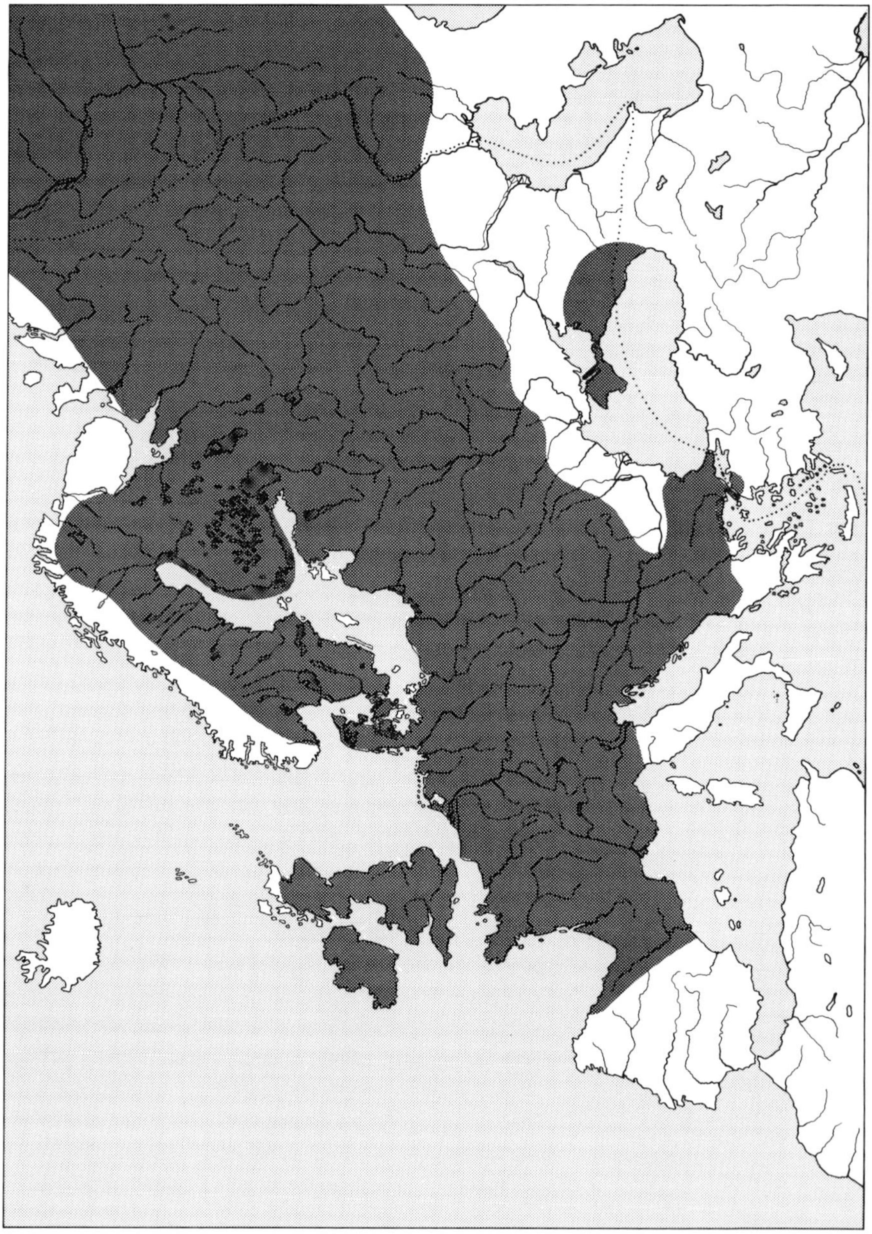

Fig. 54: European distribution of *Phoxinus phoxinus* (Linnaeus, 1758).

course, provided they are well oxygenated. This fish is never solitary; the size of its schools depends on its abundance. The schools observed recently often contain only a few (20 to 100) individuals. Spawning takes place above the stony substrate in shallow places. A number of spawning fish are often observed on the crest of a riffle. The eggs adhere firmly to the substrate. The larvae are photophobic. It is a bottom and a drift feeder. The spawning season is sometimes rather long, lasting from April to June. Maturity is reached in the second year. The life span is short; maximum 5 to 6 years. The maximum length does not exceed 12 to 13 cm. The life cycle and morphology are given in a paper by TACK (1940) but it should be noted that his observations were made in almost undisturbed natural environments, before the pressure of civilization became so great.

Population in Europe

The sizes of populations vary according to water quality and changes in the watercourse. It is abundant in unpolluted, natural streams. In polluted streams or those where there is intensive brown trout stocking, it is extremely scarce. Many populations in Central Europe are already extinct.

Reasons for decline

Pollution and man-made changes along rivers are not the only decisive reason responsible for decline. In creeks and river stretches where electrofishing is employed to remove coarse fishes and large brown trout prior to stocking with juvenile trouts the number of minnows decreases considerably. The electric current itself is apparently not lethal, but the minnow may generally be sensitive to disturbance in its biotope. The introduction of detergents was probably the main reason for its decline in smaller rivers.

Conservation measures taken

No effective conservation measures seem to have been undertaken so far. Re-introduction effort brought good results in several cases (BLESS, pers. comm.). In the Netherlands, minnow are protected by law.

Conservation measures proposed

If any effort is made to reintroduce the minnow, the environment must be suitable. The creation of artificial micro-habitats, including shelters to hide when disturbed, is a necessity. A school of fish must be released together so that they may live in a group. An intact school responds to danger with a spontaneous escape reaction.

Phoxinus percnurus (Pallas, 1911)

E. – Vernacular name:
F. –
G. Sumpfelritze Family: Cyprinidae

Classification: vulnerable to endangered

Note: Apart from the nominate form four subspecies have been described: *Phoxinus p. dybowski* (Lorec and Wolski, 1910), which occurs in the Vistula Basin, particularly in the vicinity of Warsaw; *Phoxinus p. gdaniensis* (Berg, 1932) found in the plane area round Gdynia; *Phoxinus p. posnaniensis* (Berg, 1932), described from the Oder Basin, particularly from the vicinity of Poznan, and *Phoxinus p. stagnalis* (Warpachowski, 1886), known from the lake district of central Volga.

Distinguishing characteristics

The "Sumpfelritze" is a small fish with a cylindrical body that looks basically like the minnow, but is less corpulent. The body is relatively high. Its colouring is less spectacular than that of the minnow: the dorsum is dark-brown with a green sheen; the sides are yellow with irregular small dots. It may reach a length of over 15 cm. Meristic data: D 10, A 9 to 10, 68 to 97 scales in lateral line, pharyngeal teeth in two rows.

Distribution in Europe

From the Oder System eastwards to the Kama River. The distribution depends primarily on the presence of lakes and pools with water weeds than on the boundaries of watersheds. See distribution map, p. 221.

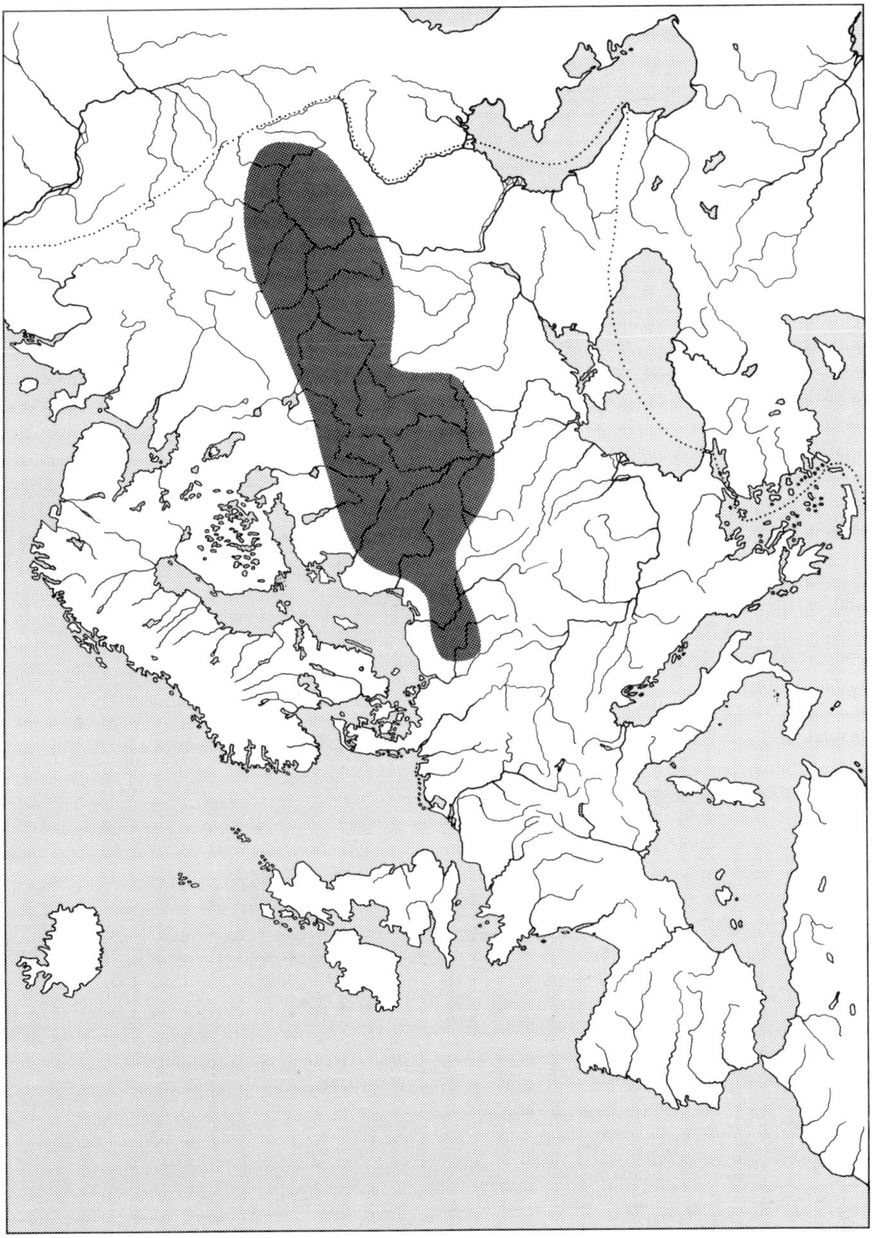

Fig. 55: Distribution area of *Phoxinus percnurus* (Pallas, 1911).

Ecology

Its German name may be translated as "Swamp minnow", well describing its habitat. It lives in closed, and usually shallow lakes overgrown with vegetation. It is not very sensitive to oxygen supply, and can survive in waters of relatively high acidity, such as pools left after peat cutting. It spawns on submerged leaves. The diet varies from pelagic to bottom invertebrates and terrestrial drift food is taken as well. It is a gregarious fish.

Population in Europe

Detailed information is not available, but communications from Poland indicate its decrease.

Reasons for decline and conservation measures

Recent efforts to drain and cultivate wetlands have been surely detrimental to this species, which requires permanent stagnant waters. Protection must be based on consequent conservation of the habitat.

Rhodeus sericeus (Pallas, 1776)

E. Bitterling
F. Bouvière
G. Bitterling

Vernacular name: –

Family: Cyprinidae

Classification: rare to vulnerable

Note: The bitterling considered here is the European subspecies *Rhodeus s. amarus* (Bloch 1782). *Rhodeus s. sericeus* (Pallas, 1776) occurs in the Amur basin and *Rhodeus s. sinensis* in the Yangtze Valley.

Distinguishing characteristics

The bitterling is a very small, deep-bodied, laterally compressed fish. Its lateral line is incomplete, extending through only the first 4 to 6 scales. The mouth is small and terminal. The colour is bright, the dorsum grey-green, the sides with a metallic streak, particularly towards the caudal body part. Over this silvery colour extends a pink sheen. This pink colour develops during spawning time, when the sides are strikingly irridescent and the fins bright red. Meristic data: D (9) 12 to 13 (15), A (10) 11 to 12 (14); 32 to 42 scales on the side, pharyngeal teeth in one row, 5 on each side.

Distribution

One distribution area of this species is in the Far East, the other in Europe. European range extends from northeastern France and the Rhône Valley over Central Europe to the Neva River. It is also found in the Black and Caspian Sea catchment areas. On the Balkan Peninsula, it occurs only in the Vardar and Maritsa Basins. It is not found south of the Alps, in northern Europe and Ireland. See distribution map, p. 224.

Ecology

The bitterling lives preferably in slow-flowing streams, backwaters and lakes, usually where the bottom is composed of fine, soft sand or covered with a thin layer of mud. However, it avoids deep muddy areas in which anaerobic conditions may occur. It requires the presence of mussels (gen. *Anodonta* or *Unio*) within the same habitat. The female lays the eggs in the gill chambers of live mussels. As a rule, it is the male which selects the host

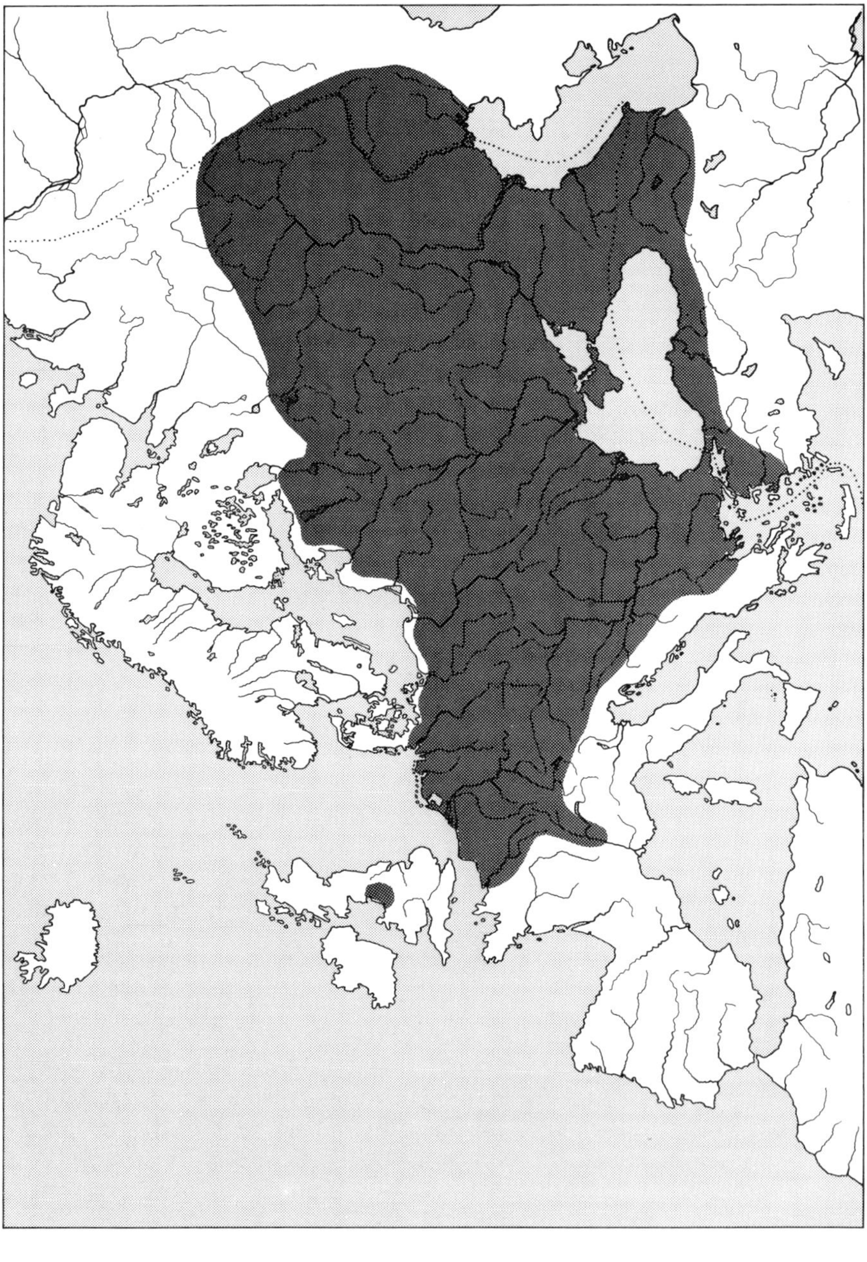

Fig. 56: Distribution of *Rhodeus sericeus amarus* (Bloch, 1782).

and ejects sperm shortly before or after the eggs have been deposited. The sperm is inhaled by the host. The eggs are deposited through a long fleshy tube from the female genital opening. The larvae remain in the mussel until they are able to feed freely: 3 to 4 weeks altogether. The bitterling feeds on phytoplankton, small benthic fauna, tubificid worms and vegetable detritus. Maturity is reached during the 2nd or 3rd year. The life-span is short, a maximum of 5 years. It has modest oxygen demands, but is very sensitive to pollution other than organic or municipal wastes.

Population in Europe

The population in Europe is rather unstable. In many areas of Central Europe formerly well populated, such as Rhine-Main Region, it has completely disappeared from many localities.

Reasons for decline

There is no doubt that pollution, especially industrial pollution, is responsible for the decline of this species. In many localities, mussels survive but bitterlings have disappeared, e.g. the central reaches of the Rhine. It is not known at which stage of life this fish is most sensitive to pollution.

Conservation measures taken

The protection of certain backwaters and oxbows which do not communicate with the main water course has rescued and conserved some populations. In some countries, the bitterling is protected by law, for example in the Netherlands. Very encouraging results were achieved by means of rearing in special ponds in Germany.

Conservation measures proposed

Pollution control, protection of solitary habitats and re-introduction could improve the diminishing stock of this species.

Rutilus pigus pigus (Lacépède, 1804)

E. –
F. –
G. –

Vernacular name: Pigo (Italian)

Family: Cyprinidae

Classification: rare

Note: In the upper part of the Danube *Rutilus pigus* is represented by the subspecies, *Rutilus pigus virgo* (Heckel, 1852). As this subspecies is declining in Central Europe, it will be treated in detail separately, p. 227.

Distinguishing characteristics

Rutilus pigus looks like a long-bodied roach *(Rutilus rutilus),* but it is not as high, the sides are more compressed laterally, and the mouth is not exactly terminal but slightly inferior. The colouring is the most conspicuous pattern of this species. The dorsal part is dark green, the sides of the body are a shiny blue-green. During spawning time, white tubercles develop on the head and extend caudally over the dorsum. Meristic data: D 12 to 14, A 14, the pharyngeal teeth are well developed in one row, 6 (5) – 5.

Distribution in Europe

The Pigo is found in northern Italy, particularly in Lakes Maggiore, Lugano, Como, and Garda and in stretches of the Rivers Po, Adige, Piave and perhaps some others. See distribution map, common for *Rutilus p. pigus* and *Rutilus p. virgo,* p. 228.

Ecology

Its ecology is little known. It lives preferably in deep waters, feeds on molluscs, crustacea and worms, spawns in shallow water, and may reach a length of 40 cm.

Population in Europe

Very little is known of its abundance, as it attracts very little commercial or sport fishing interest.

Reasons for decline

A lack of data prevents any determination of decline.

Conservation measures taken or proposed

None known, but the life-cycle of this species should be investigated, so that effective conservation measures can be suggested. On ground of the small range, inhabited by the Pigo, protection is surely recommended.

Rutilus pigus virgo (Heckel, 1852)

E. Danubian roach — Vernacular name: –
F. –
G. Frauennerfling — Family: Cyprinidae

Classification: rare and vulnerable

Distinguishing characteristics

The "Frauennerfling" resembles a long-bodied roach, and its colouring is similar to that of *Rutilus p. pigus*. This fish is very conspicuous during spawning time, when white spawning tubercles extend from the head over the dorsal part of the body, and on the sides, extending to the lateral line. The position of the mouth is semi-inferior; the lips relatively thick. The anal and pelvic fins are orange-red, the pectoral yellow-white, the belly white with a little blue shading. The lining of the body cavity is dark. Meristic features: D 12 to 14, D 13 to 16, 44 to 49 scales in and 7 rows of scales above the lateral line. Pharyngeal teeth well developed, 6 to 5 or 5 to 5.

Distribution in Europe

This fish is found only in the Danube and the lower stretches of its larger tributaries, from Bavaria downstream to Hungary. It was recorded in northern Romania in tributaries of the Tisza River. See distribution map, p. 228.

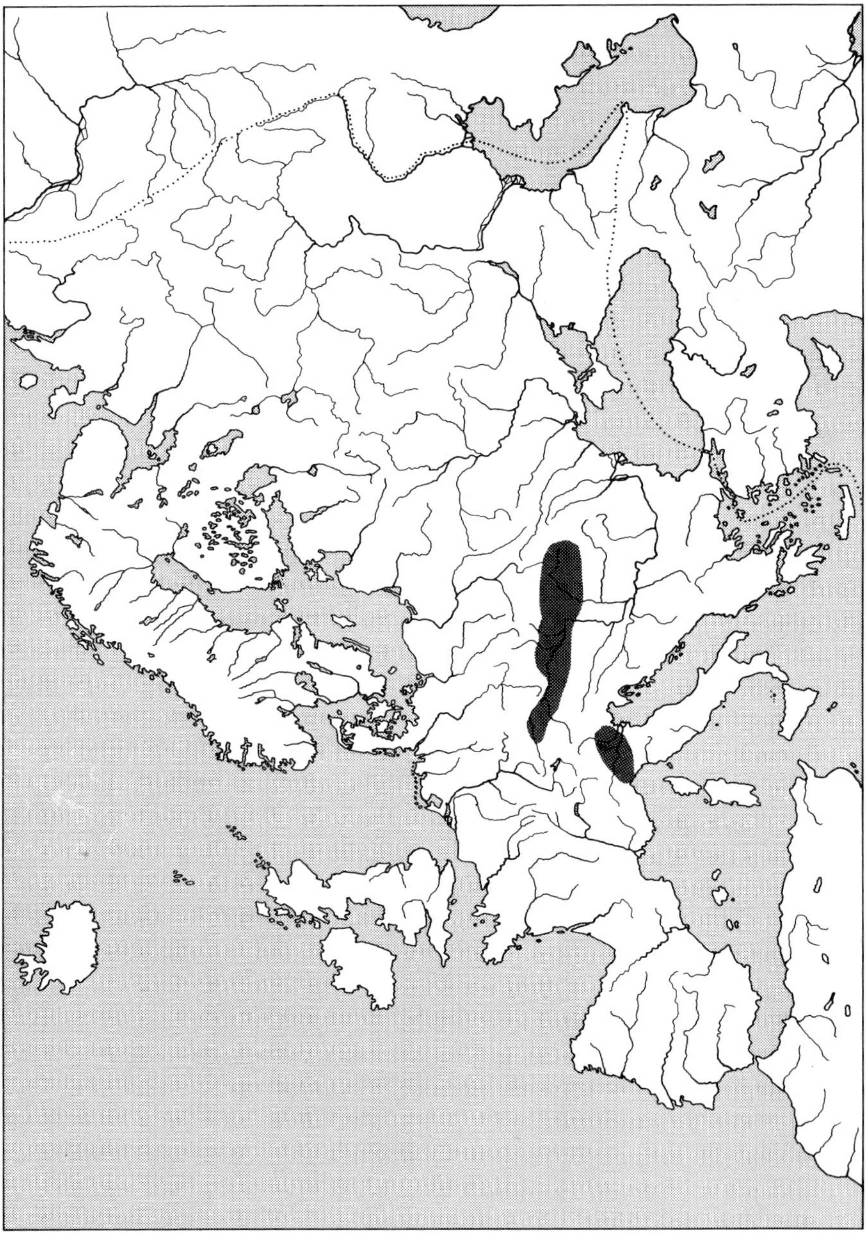

Fig. 57: Distribution areas of *Rutilis pigus pigus* Lacépède, 1804 (River Po) and *Rutilis p. virgo* (Heckel, 1852) – Danube area.

Ecology

This species usually lives in the Danube itself or its larger tributaries, at depths with a moderate current. Only during the spawning time (April to May) does it enter tributaries and backwaters with dense plant growth. It is a phytophilous spawner. Maturity is reached in the 3rd year of life and it may reach a length of 40 cm and a weight of 1 kg. The maximum life-span is 15 years. Food and feeding habits are not known.

Population in Europe

Its scarcity and mode of life make it difficult to obtain information about its former and present abundance to evaluate its decline. However, there is some indication, that it is now rarer in the Bavarian Danube and some tributaries in Moravia and Slovakia than it was several years ago.

Reasons for decline

Not known exactly. It is assumed that this subspecies is sensitive to pollution, environmental changes and construction works, that have changed the velocity of the currents in many places particularly in the Danube.

Conservation measures taken and proposed

No special conservation measures are known as this subspecies has little value as a commercial or sport fish but it is now legally protected in Central Europe. It is hoped that the general rules for conservation of the water-bodies adjacent to large rivers, as mentioned in section 4, will help this species to survive.

Rutilus rubilio (Bonaparte, 1837)

E. –

F. Gardon de l'Europe du Sud

G. Plötze (Swiss)

Vernacular name: Triotto (Italian)

Family: Cyprinidae

Classification: intermediate

Note: Two subspecies from the Balkan Peninsula have been described: *Rutilus rubilio ohridanus* (Karaman, 1924) occuring in Lake Ohrid and *Rutilus r. prespensis* (Karaman, 1924) from Lake Prespa in Macedonia.

Distinguishing characteristics

This species has the general roach habitus with a relatively high body. The length of the fins is important for identification: Both the dorsal and the caudal fins are longer than wide, both have a short base. The lateral line is bent towards the belly. On the silvery background of the sides, metal-grey stripes are sometimes clearly visible. Meristic data: D 12 to 14, A 11 to 14, pharyngeal teeth in one row, 5 to 5.

Distribution in Europe

This species occurs from the Tagliamento, Adige and Po River valleys to southern Italy. To the east it is found in the rivers emptying into the Adriatic Sea. The most eastern occurrence is the basin of Aspropotamos and both sides of the Dardanelles. See distribution map, p. 231.

Population in Europe

Exact data are not available, but in many places it is very common.

Ecology

This is a gregarious fish, found mostly in stretches free from plants. Spawning takes place in the litoral zones among water plants. It takes plankton, benthic and drifting food.

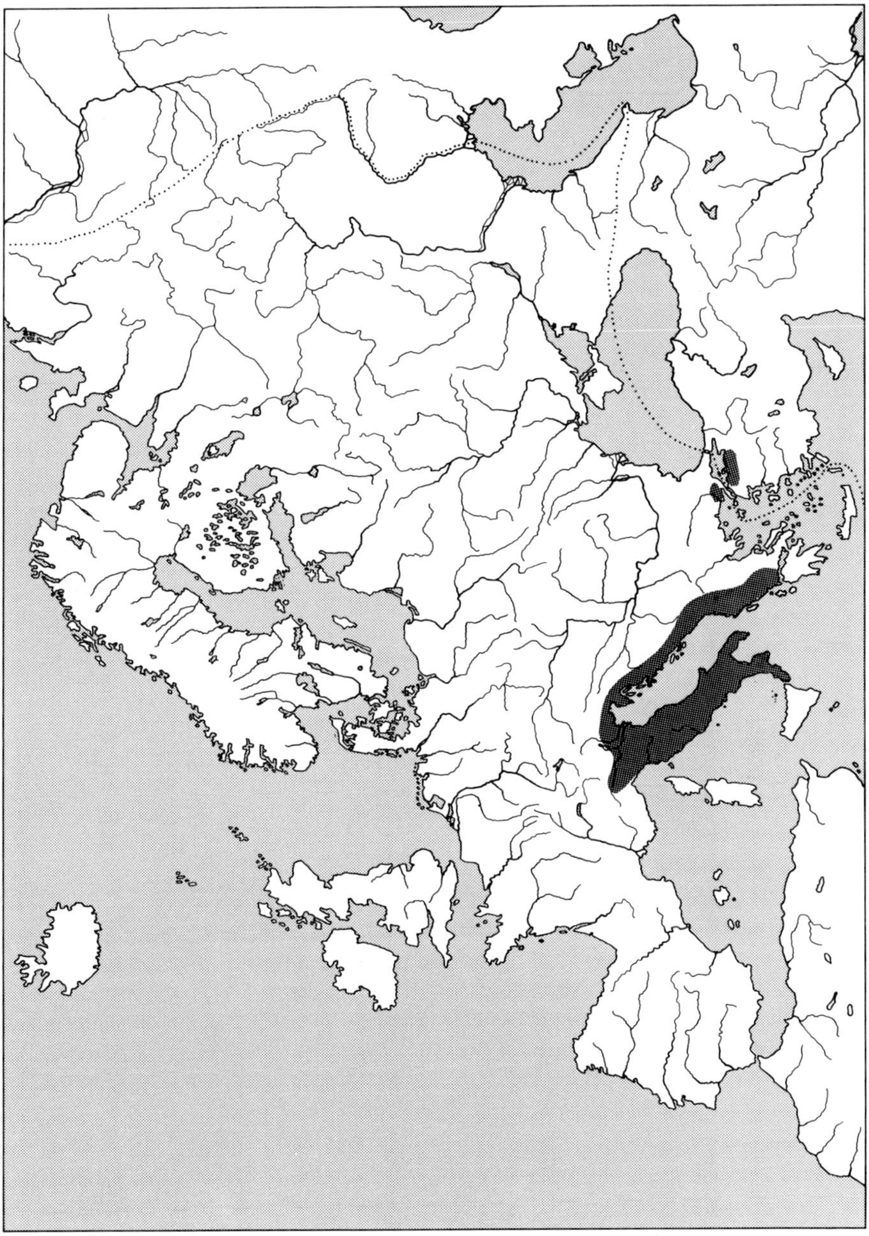

Fig. 58: Distribution of *Rutilus rubilio* (Bonaparte, 1837).

Reasons for decline

No decline has been reported but that may be due to the fact, that no reliable informations are available on former and recent abundance. It is assumed that changes of shore-line may be detrimental to this species.

Conservation measures taken and proposed

Non conservation measures are known. Because this species is fished commercially at some places, restrictions of catches would be advisable.

Pararutilus frisii meidingeri (Heckel, 1852)

E. – Vernacular name: Frauenfisch (German)

F. –

G. Perlfisch Family: Cyprinidae

Classification: rare and in great danger

Note: Sometimes named as *Rutilus* or *Leuciscus f. meidingeri.*

Distinguishing characteristics

The "Perlfisch" is a long-bodied, robust-built cyprinid. The cross-section of the body is almost round, the small mouth terminal. The colour is dark green; the sides brighter, turning to silver, the belly is silver-white, the fins grey and transparent. During spawning, it has strongly developed tubercles on the body, more numerous above than below the lateral line, and none on the belly. Meristic data: D 11 to 12, A 12 to 14, 62 to 67 scales in the lateral line.

Distribution in Europe

This fish is native to several Subalpine lakes: Chiemsee, Traunsee, Mondsee, and adjacent streams. Dubious reports from the Danube and its tributaries probably resulted from confusion with *Rutilus pigus virgo*. See distribution map of *Pararutilus frisii,* p. 234.

Ecology

This is a lake fish, living almost constantly in small schools in deep water. It spawns during spring (April and May) in streams connected to lakes and their outlets. It feeds on invertebrates, small fish, and even vegetation. It may reach a length of over 60 cm, and its life expectancy is over 12 years.

Population in Europe

Very little is known about the present stock.

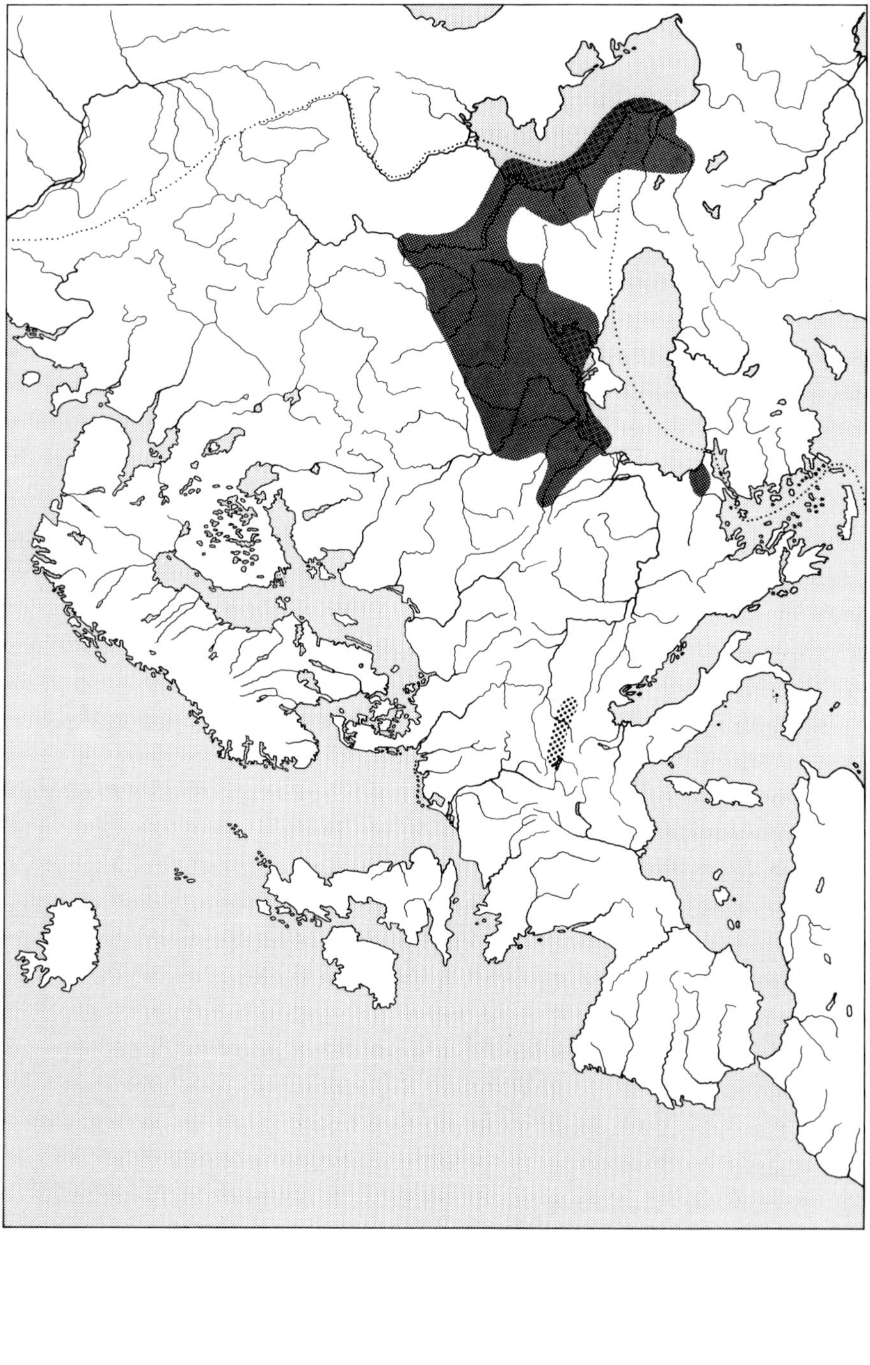

Fig. 59: Distribution of *Pararutilus frisii* (Nordmann, 1840) ▩ with subspecies *Pararutilus f. meidingeri* (Heckel, 1852) ⁘.

Conservation measures taken or proposed

No actual conservation measures have been recorded, and it is hoped it will not be adversely affected by eutrophication. Fishing in the tributaries during spawning time is already forbidden in some places. But legal protection, as recently enacted in southern Germany, seems to have little effect without consequent and effective protection of spawning grounds. It is hoped that this will soon be achieved.

Scardinius erythrophthalmus (Linnaeus, 1758)

E. Rudd
F. Rotengle
G. Rotfeder

Vernacular name: –

Family: Cyprinidae

Classification: vulnerable

Note: Apart form the nominate species, two subspecies are recognised at present: *Scardinius e. ohridanus* Vlad. from Lake Ohrid and *Scardinius e. scardafa* (Bonaparte, 1832) from northern and central Italy, including the Tiber River Basin, and southeastern Yugoslavian and Albanian rivers draining into the Adriatic. Closely related to *S. erythrophthalmus* are the two endemic species *Scardinius graecus* Stephanidis, 1937 and *Scardinius racovitzai* G. Müller, 1958.

Distinguishing characteristics

The rudd looks much like the roach *(Rutilus rutilus)*. The following characteristics can be used to avoid misidentification. The mouth is almost superior; the dorsal fin originates well behind the base of the pelvic fins. A sharp scaly keel runs along the belly between the pelvic and anal fins. The colouring is variable: the dorsal part dark, the sides brassy yellow, the belly almost white. The dorsal and caudal fins are greenish-brown with a red tinge, the pectoral, pelvic and anal fins are blood red. Meristic data: D 10 to 13, A 12 to 15, 37 to 43 scales in the lateral line and 7 to 8 rows of scales above it; pharyngeal teeth in 2 rows, 5 plus 3 on each side.

Distribution in Europe

The rudd occurs northward from the Pyrenees and across Europe to the Ural and the Aral Sea Basin. It is absent from northern Scotland, western Norway, central and southern Sweden, and northern Finland. The southern range limit probably is in central Italy. It is found in the Adriatic watershed, on the Balkan Peninsula, in southern Turkey and eastward to the Caspian Sea.

Ecology

The rudd prefers clean oxbows along larger rivers with a large, open water surface and vegetation on the banks. It also occurs in slow-flowing streams with aquatic vegetation. It matures in its second, or more usually in its third year and spawns late in spring on submerged water plants, in small

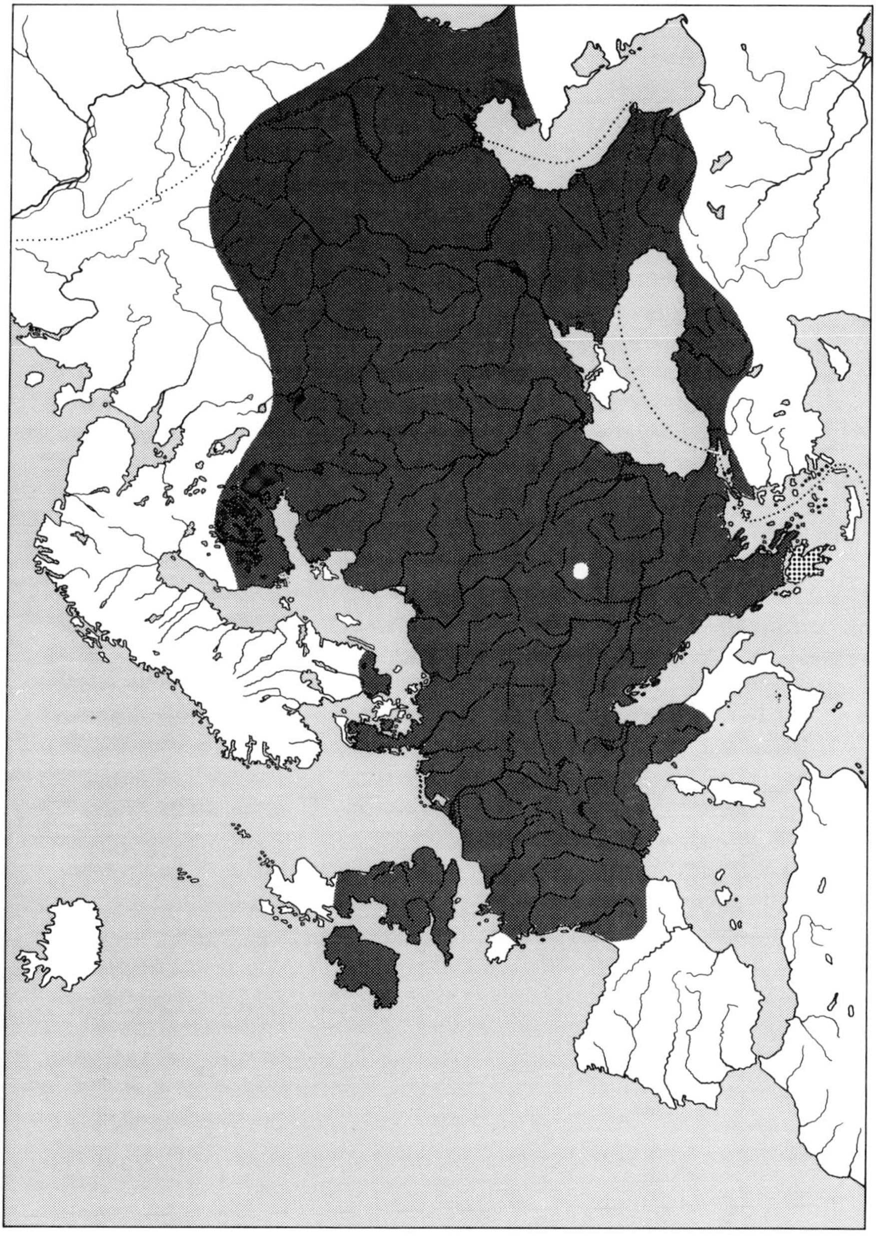

Fig. 60: Distribution of *Scardinius erythrophthalmus* (Linnaeus, 1758), including subspecies, and distribution of endemic species *Scardinius graecus* Stephanidis 1937 (dotted area) and *Scardinius racovitzai* G. Müller, 1958 (white spot N of the Danube).

groups or perhaps in pairs. The eggs are deposited in two batches. The life expectancy is rather high, over 10 years. Its food is somewhat specialised: In the early life stages zooplankton is preferred, but later the rudd lives almost entirely on algae and higher plants, only occasionally taking invertebrates, as well. It can withstand almost complete anaerobic conditions and high water temperatures, but seems sensitive to pollutants of all types, especially industrial, inorganic waste waters.

Population in Europe

It is a common fish, but very seldom a dominant species. Though found in almost all slow-flowing streams, lakes, and adjacent pools, it is always low in number of individuals.

Reasons for decline

Straightening of rivers and destruction of shore vegetation, construction of steep banks, and consequent lack of submerged plants have resulted in extermination of this species in long river stretches. Where industrial pollution has also lowered the water quality, *Scardinius erythrophthalmus* survives only in scattered backwaters, some lakes, and isolated permanent pools.

Conservative measures taken

To increase the abundance of this species, attempts are being made to introduce it into its former habitats and into small water bodies left after the excavation of sand and gravel. During attempts to release rudd into waters dedicated to angling, the roach is often introduced instead, as confusion of these two species is frequent. The result is overstocking with roach, which has no special environmental demands. The roach spawns much earlier, and the larvae of the less abundant rudd that hatch later, are then eaten by the crowded roach population. No legal protective measures are known, and it is doubtful whether they would be successful.

Conservation measures proposed

Protection of oxbows and clean, old water bodies would best ensure the conservation of this species. Stocking should be supervised by trained fishery managers, to avoid the introduction of the already abundant roaches.

Note: Proper use of this indigenous species might prove more effective in algae and soft-weed control than attempts to introduce herbivorous Asian fishes, such as the grass-carp *(Ctenopharyngodon idella)*.

Tinca tinca (Linnaeus, 1758)

E. Tench
F. Tanche
G. Schleie

Vernacular name: –

Family: Cyprinidae

Classification: intermediate

Distingushing characteristics

The tench has a thick-set and heavy body proportion. The body is covered with very small scales, which are deeply imbedded in the skin. The mouth is terminal, small and has a single barbel at each corner. Both dorsal and anal fins are short based, with not more than nine branched rays. All fins are rounded. Colouring varies with locality, but is basically dark to black-green on the dorsum with a golden sheen on the sides, bronze to yellow on the belly. The tench displays a pronounced sexual dimorphism: the male has thick pelvic fins. Meristic data: D 11 to 12, A 9 to 11; 87 to 115 scales in, and 30 to 35 rows of scales above the lateral line.

Distribution in Europe

The tench is found throughout Europe with the exception of Iceland, the northernmost part of Scotland, Norway, the central and northern parts of Sweden and Finland. Farther northeast, it does not occur north of 61° N. In the South, it is absent from rivers flowing to the Adriatic Sea from eastern Yugoslavia, the Balkan Peninsula, Greece and the Crimea. It has been introduced to Sardinia. It occurs in the Black, Caspian and Aral Sea Basins (see distribution map). It has also been introduced to North America.

Population in Europe

A first glance at its range suggests that the tench is in no danger. But this assumption proves to be wrong, when the situation is examinated carefully. The tench is an intensively cultivated fish, stocked into almost all water bodies. But natural reproduction and independent populations are found only in very few suitable water bodies.

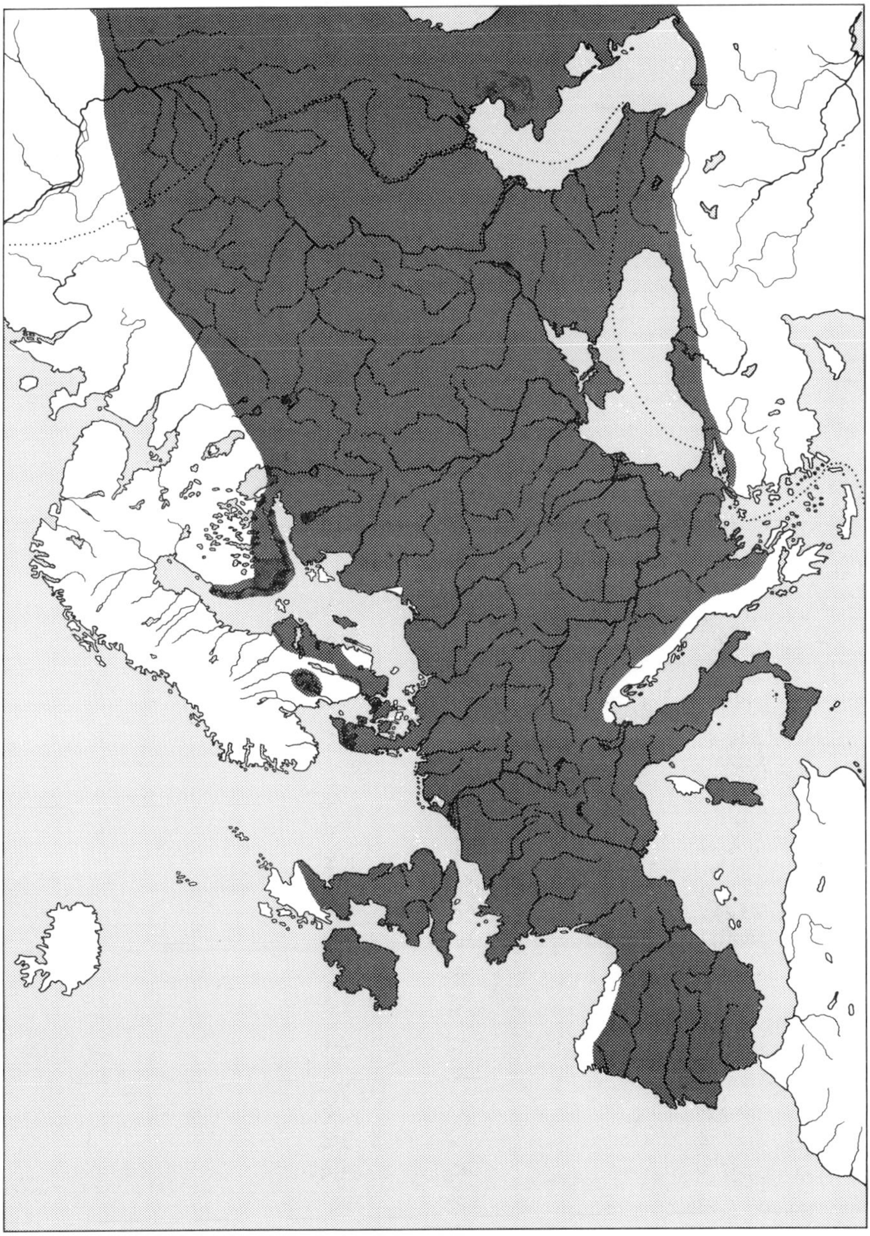

Fig. 61: European distribution of *Tinca tinca* (Linnaeus, 1758).

Ecology

The tench generally lives in lentic waters, such as lakes, the lower reaches of rivers and adjacent backwaters, oxbows and marshy deltas. It can tolerate semisaline conditions. Its ability to survive in waters almost without oxygen, to tolerate high summer temperatures and to survive the winter under almost anaerobic conditions makes *Tinca tinca* relatively insensitive to physical conditions of the water. Spawning takes place very late, usually at the beginning of summer. The eggs are always deposited on vegetation, which need not necessarily be composed of soft leaves. The growth rate reflects more the suitability of the environment than the age. The tench may grow very large (up to 50 cm) and reach an old age. It is not a specialized feeder, preferring benthic invertebrates, but also feeding on molluscs, vegetation, and detritus.

Reasons for decline

Decline of natural populations is often not observed, as it is compensated by stocking.

Conservation measures taken

In Central Europe, the tench is commercially important, so legal size limits and closed seasons have been introduced in many countries. Stocking, practised mostly by angling clubs, contributes to its present distribution and abundance.

Conservation measures proposed

Protection of the lower stretches of rivers, the extension of shoreline vegetation, and connecting backwaters to the main stream, can all help this species to be less dependent on artificial stocking.

Vimba vimba (Linnaeus, 1758)

E. –
F. –
G. Zährte

Vernacular name: –

Family: Cyprinidae

Classification: intermediate and rare

Note: In the genus *Vimba,* three species and four subspecies are sometimes distinguished, or it is considered to include only one species with 6 Subspecies (Ladiges and Vogt, 1979), which opinion is followed by us.
Vimba v. bergi (Velikokhatko, 1940) occurs in the lower reaches of the River Bug.
Vimba v. carinata (Pallas, 1811), recently considered as synonymous with *Vimba v. vimba,* found from the Danube to the Kuban (Bănărescu, 1970).
Vimba v. elongata (Valenciennes, 1844) lives in the natural lakes of southern Bavaria and Austria.
Vimba v. melanops (Heckel & Kner, 1858) occurs in several rivers on the Balkan Peninsula (Maritsa, Mesta, Struma, Vardar).
Vimba v. persa (Pallas, 1811) occurs in the lower reaches of the Volga and in the Caspian Sea.
Vimba v. tenella (Nordmann, 1840) lives in small rivers emptying into the Black Sea, e.g. the Krim and the Kuban, and in western Transcaucasia.

Distinguishing characteristics

Vimba has a moderately deep, laterally compressed body, and a scaleless keel on the belly between the anal and the caudal fins. On the dorsum, there is a keel covered with scales, from the dorsal fin to the tail. The mouth is inferior with thick, horse-shoe shaped lips. The snout is fleshy and projects further than the diameter of the eye. The colour on the dorsum is blue-grey, during spawning almost black. It is yellowish-silver on the belly. During the spawning period, the insertion of the paired fins, the area of the gill cover and often the whole underside including the anal fins become orange red. Meristic data: D 11 to 12, A 20 to 25; 51 to 64 scales on 9 to 10 scales above it; pharyngeal teeth in one row, 5 on each side.

Distribution in Europe

This species is absent from Western Europe, its distribution beginning east of the Elbe River. The Weser River is sometimes considered as the westernmost limit of range. In Scandinavia, it occurs only in southern Sweden and Finland. In the South-east, distribution includes the Danube River

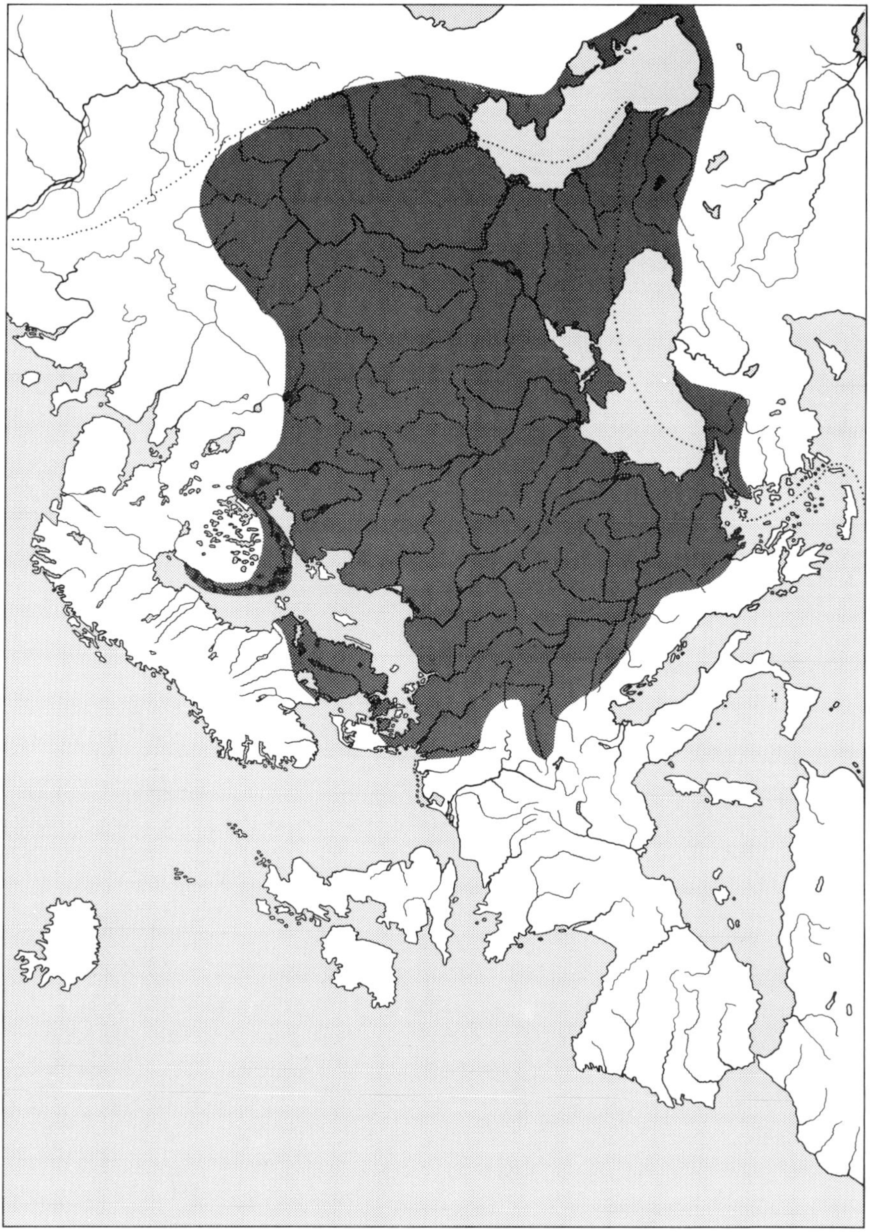

Fig. 62: Distribution of *Vimba vimba* (Linnaeus, 1758).

System. In the East it is found from the Vardar Basin on the Balkan Peninsula, to western Turkey and further to the Volga, Ural and Emba River Systems, see the map, p. 244. With stocking introduced in the Rhine and Mosel Rivers (about 1983).

Ecology

All subspecies of *Vimba vimba* are found in the middle zones of rivers. During the spawning season it ascends into the zones occupied by *Chondrostoma,* often migrating for long distances. Spawning takes place on a clean river bed, free from mud or silt deposits. After releasing the eggs, in several, usually three, batches, it returns downstream below the barbel zone and often as far as the bream zone. It often occurs in the mouth of the rivers leading to the Baltic Sea. Maturity is reached in the 3rd or 4th year of life. It is reasonably adaptable, living in man-made lakes and spawning in rivers, feeding those. Often is observed the occurrence of hybrids, particularly with *Blicca bjoerkna.* Populations living in the Subalpine lakes of Bavaria and Austria have similar habits. It is a bottom feeder; HOLČÍK and HENSEL (1972) noticed a trend to further westward expansion in Slovakia.

Population in Europe

As a general rule, the stock of this species is larger in its eastern distribution area than in its western one, but it has never been very abundant. During spawning the impression of greater abundancy may be produced by concentration of individuals in those regions suitable for spawning. Its occurrence in the Weser and Mosel (Germany) may be caused by stocking with fish from the Danube drainage area.

Reasons for decline

Decline has been observed only in north-western part of its range. In the Elbe, the populations decline due to the pollution with industrial wastes, though *V. vimba* is not particularly sensitive to low oxygen supply. In the upper Danube a frequent hybridization with the white bream *Blicca bjoerkna* may be the reason for its decline (Balon et al. 1987).

Conservation measures taken or proposed

No special conservation measures are known. It is hoped that a decrease in pollution and better access to spawning places might solve the conservation problems.

3.13. Cobitidae

Cobitis taenia Linnaeus, 1758

E. Spined loach
F. Loche de rivière
G. Steinbeißer

Vernacular name: –

Family: Cobitidae

Classification: rare

Note: This widely distributed species is known to have several subspecies:
Cobitis t. bilineata (Canestrini, 1866), found in northern Italy;
Cobitis t. dalmatina (Karaman, 1928) lives in the river Cetina, Dalmatia;
Cobitis t. haasi (Klausewitz, 1952) occurs in eastern Spain;
Cobitis t. meridionalis (Karaman, 1924) found only in Lake Prespa;
Cobitis t. paludicola (F. de Buen, 1930) occurs in the Tejo Basin and on the most northern tip of Africa;
Cobitis t. puta (Cantoni, 1882), found in the Po, Brenta and Dese Basins, northern Italy;
Cobitis t. strumicae (Karaman, 1955) lives in the Struma Basin and probably in the Mesta and Maritsa Rivers;
Cobitis t. zanandreai (Caricchioli, 1965) lives in Campania, Italy;
Cobitis t. calderoni Bacescu, 1961, occurs in northern Spain and is often referred to as a species.

Distinguishing characteristics

The spined loach has the general loach habitus, with an elongated laterally compressed body, covered with minute scales. The head and mouth are small. It possesses 6 barbels, four at the snout tip and two at the corners of the mouth. Below the eye, there is a strong double-pointed, retractible spine, pointing caudal. The second ray in the pectoral fin is thicker in the males and bears a rounded shiled called Canestrini's scale close to its insertion. There are 10 to 20 rounded blotches in a clearly visible longitudinal row along the sides. On the dorsum, at the insertion of the caudal fin, there is a black, or at least a very dark spot. Meristic data: D 8 to 10, A 7 to 9.

Distribution in Europe

The spined loach including all its subspecies, is distributed throughout most of Europe. It is not found in Iceland, Ireland and a great part of Eng-

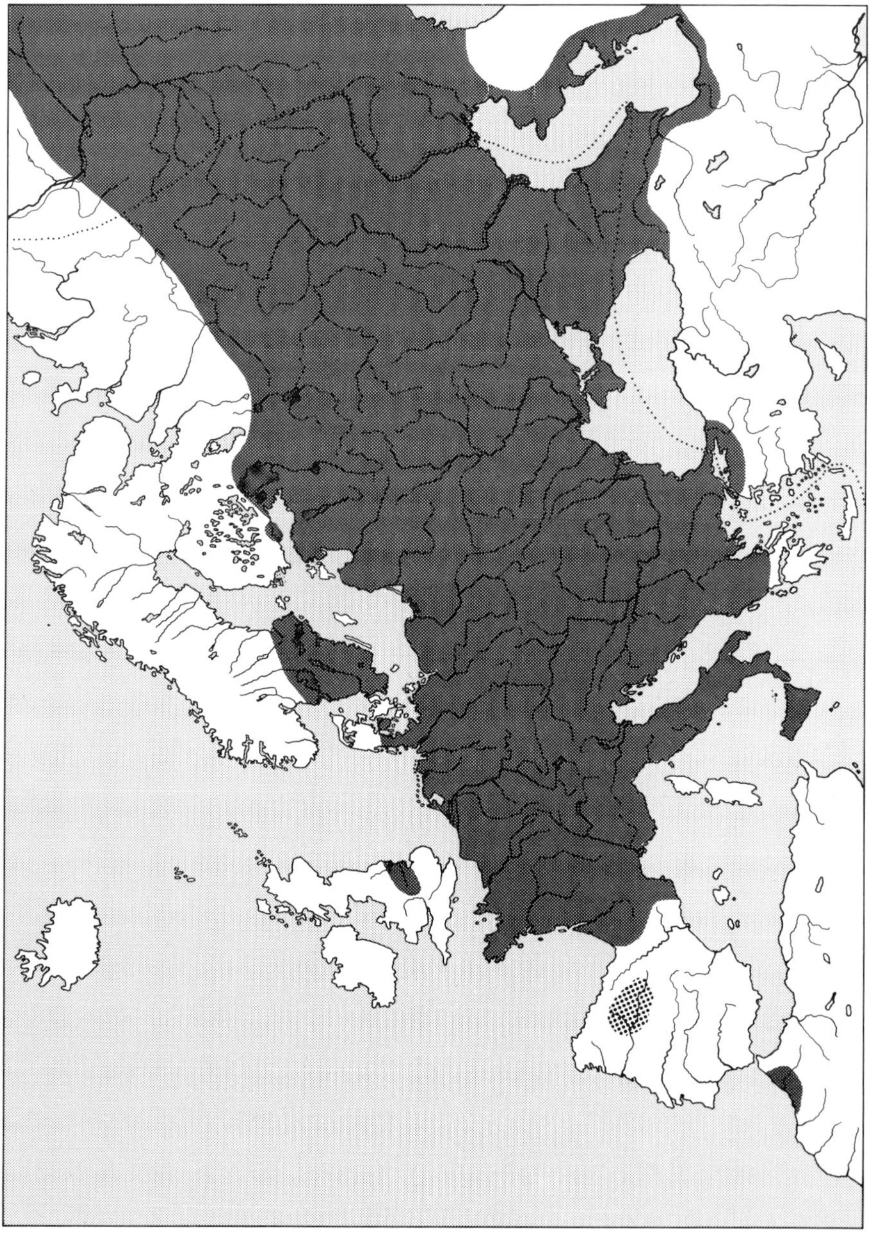

Fig. 63: General Distribution of *Cobitis taenia* Linnaeus, 1758 (including subspecies) in Europe. Dotted area: *Cobitis t. calderoni* Bacescu, 1961.

land, Norway, Sweden north of the 60th parallel. It is also absent from most of Finland and the northernmost areas of Europe and Asia. Its range ends at the Lena River Basin. In the south, it occurs in the northernmost part of Africa, Italy including Sicily, in the Vardar and Maritsa Basins and throughout Dalmatia. It is absent from the southern tip of Greece and Turkey, with the exception of the streams in the western part of the country, adjacent to the Bosporus-Dardanelles strait.

Ecology

Ecology of *Cobitis taenia* is little known as it lives buried in sand during the daytime. It occurs in small creeks with sandy bottoms, as well as in lowland lakes and channels, provided the water is clean and the layer of mud is not thick or coherent. Its activity begins at twilight and continues during the night. It prefers cool waters. It is never abundant, and it seems that individuals remain usually solitary. Spawning time extends from May to June. The eggs are laid on the sand, roots of floating plants and submerge vegetation. The maximum size of adult individuals does not exceed 10 to 12 cm. It is believed to feed on invertebrates as well as vegetable matter, but no detailed investigation of food and feeding habits have been made.

Population in Europe

No detailed information is available on the size of its populations. According to electro-fishing data, it is more common in clean waters, than expected, but is never dominant.

Reasons for decline

Very little is known about its abundance, so it is impossible to state the trends of its decline or increase.

Conservation measures taken and proposed

No conservation measures are known, except legal protection in some countries, such as Netherlands. It was, however, observed that the straightening of small streams and reinforcement of the shore-line did not affect the population adversely. But these observations were made only in rivers with sandy beds. In habitats with less suitable bottom human influences may cause threat to the populations. Pollution control may be the most effective measure for protecting this species. The predation by eels cannot be ruled out.

Cobitis (Sabanejewia) aurata (Filippi, 1856)

E. Goldside loach? Vernacular name: –
F. –
G. Goldsteinbeisser Family: Cobitidae

Classification: rare to vulnerable

Note: At least four subspecies according to Ladiges and Vogt (1979) are recognised at present:

Cobitis (Sabanejewia) a. balcanica (Karaman, 1922) in Danube tributaries from the River March (Czechoslovakia) eastward; in the systems of the Vardar, Struma, Mesta and Maritsa Rivers on the Balkan;
Cobitis (Sabanejewia) a. bulgarica (Drensky, 1928) in the central and lower reaches of the Danube and in its tributaries.
Cobitis (Sabanejewia) a. radnensis (Jasfalusi, 1951) is a local form restricted to the upper reaches of the River Muresul in northwest Romania.
Cobitis (Sabanejewia) a. vallachia (Nalbant, 1957) occurs in Danube tributaries in southeast Romania such as the River Seret.

Dinstinguishing characteristics

This small species has the elongated body, typical of a loach, covered with minute scales. Below the middle of the eye there is a retractable, double-pointed spine. The small mouth is inferior, with six barbels around it. The colouring is rather conspicuous: the brownish-olive green dorsal part has a dark marble pattern; the sides have a golden-brown shine with a row of dark-brown spots. It may reach a length of 10 to 12 cm. Meristic data: D 8 to 10, A 7 to 9.

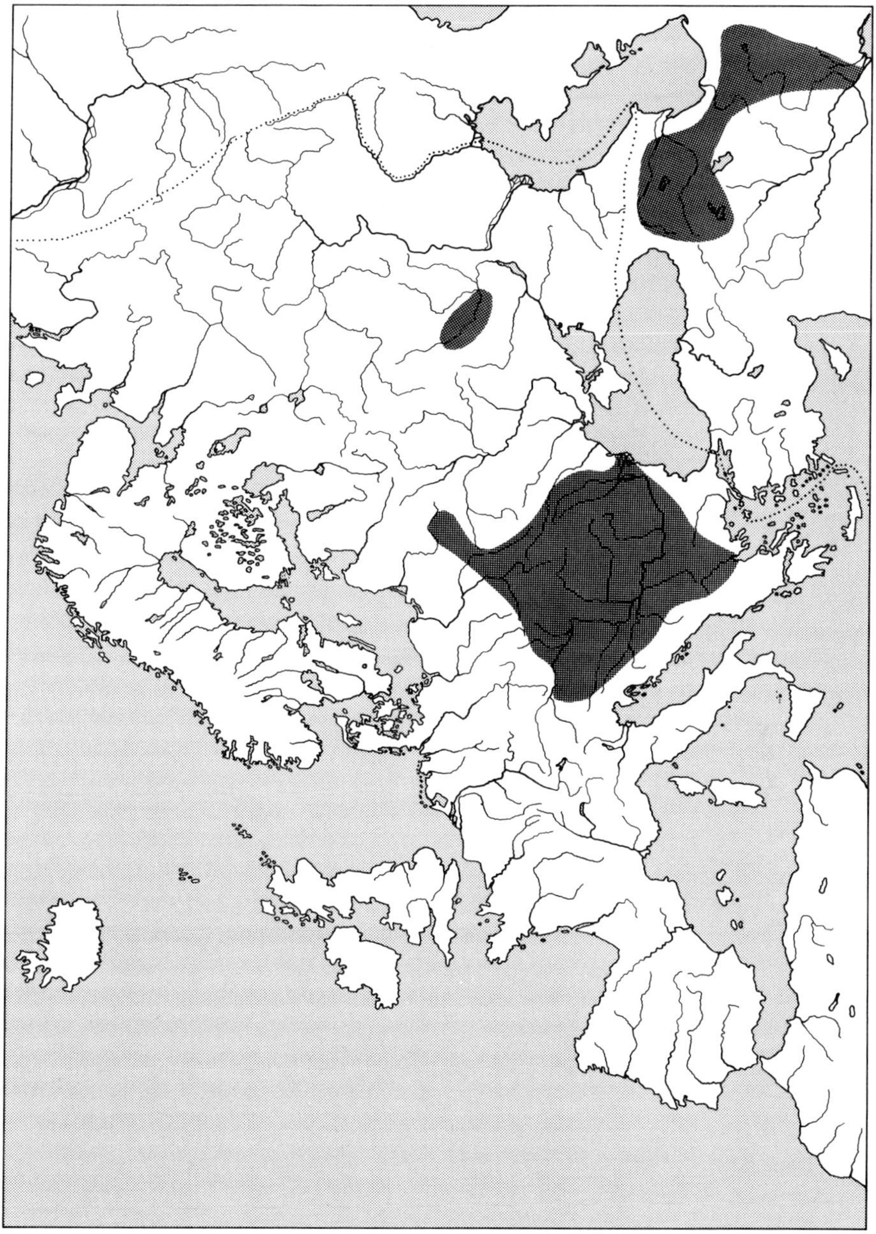

Fig. 64: Distribution of *Cobitis aurata* (Filippi, 1856).

Distribution in Europe

The distribution area of this species is not fully known, because it lives hidden. The information given here may reflect only localities of intensive sampling. However, it can be stated, that the actual main range in Europe is confined to the Danube Basin, some Balkan and Caucasian rivers, and from the Baltic watershed to the upper reaches of the Vistula, Neman and Sanie Rivers, see distribution map, p. 251.

Ecology

Cobitis aurata is mainly nocturnal, and during the day it stays buried in the sandy bottom. It occurs in flowing stretches of rivers; its presence in muddy or silted places is an exception. It is a phytophylic spawner and does not protect the eggs. The spawning time extends from April to June but usually takes place in May. It is supposedly a bottom feeder, but neither its ecology nor its environmental requirements are known in detail. It may survive in waters that are rather warm during summer, slightly over 20° C.

Population in Europe

Because of its secretive life habits and the difficulty to distinguish this species from *C. taenia*, the size of its population is unknown. It is common in unpolluted Danube tributaries in the central part of its distribution area.

Reasons for decline

The fish disappeared in some places only after large man-made environmental alterations, but its sensitivity to human activities is not accurately known.

Conservation measures taken and proposed

No conservation measures have been taken, and none can be proposed owing the lack of information.

Misgurnus fossilis (Linnaeus, 1758)

E. Weatherfish
F. Loche d'étang
G. Schlammpeitzger

Vernacular name: –

Family: Cobitidae

Classification: rare and vulnerable

Distinguishing characteristics

The weatherfish has a typical loach body habitus with an elongated, almost cylindrical body, laterally compressed in the caudal region. The inferior mouth has 10 barbels, four on the upper jaw, two at the corners of the mouth and four on the lower jaw. The dorsal fin is located behind the mid-point of the body. The scales are minute, embedded deeply in the skin, and the whole body is covered with a thick layer of slime. Colouring varies according to the locality, from dark brown to red-brown on the dorsal side. Along the lateral line there is a longitudinal stripe of the same colour as the dorsum. There are two other light and dark stripes on the sides that are sometimes not clearly visible. The belly is yellow-brown, turning to bright-orange during the spawning period. The fins have small, dark, irregular spots. Meristic data: D (7) 8 to 10 (11), A 8 to 9 (11).

Distribution in Europe

This species is found from northwestern France eastward to the Volga Basin. It is present in the Danube catchment area. It is not found in Scandinavia or eastward of the Neva River. In the Southwest it is absent from the valleys of the Garonne, the Rhône, and all other rivers emptying into the Mediterranean, Adriatic and Aegean Seas. See distribution map, p. 254.

Ecology

The weatherfish is one of the best adapted fishes to life under extreme conditions, including occasional oxygen deficiency. Its typical habitat are oxbows, pools and periodically flooded marshes adjacent to large rivers, and muddy lowland rivers where the current is negligible. Adults can utilize atmospheric oxygen by swallowing air bubbles and absorbing the oxygen through a special gut epithelium. The larvae have well developed external

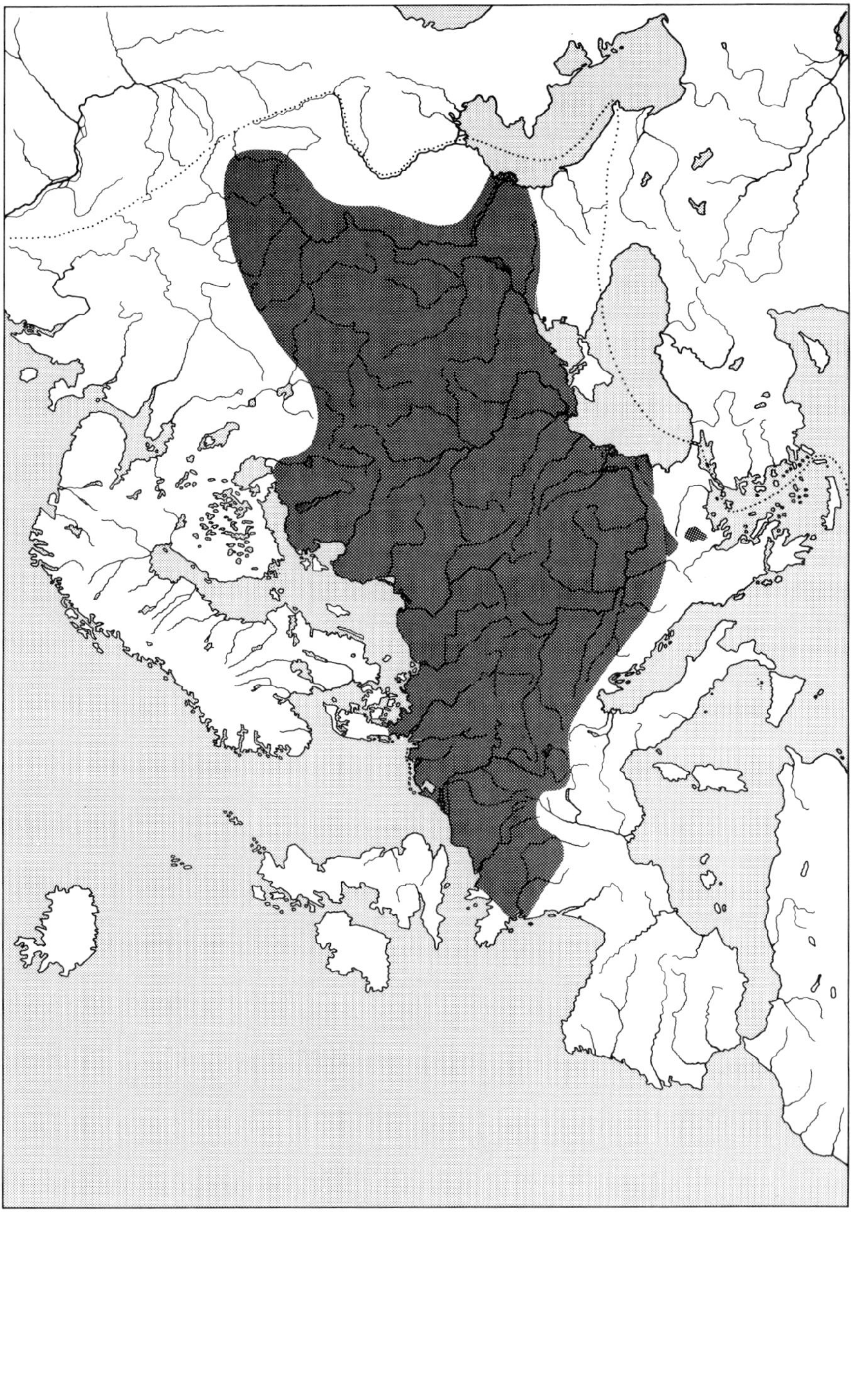

Fig. 65: Distribution area of *Misgurnus fossilis* (Linnaeus, 1758).

gills. During the day, the fish lies buried in mud; its activity reaches its peak at late twilight and continues during the night. It can withstand high water temperatures, and it is thought to survive for some time in mud. It is a phytophilous spawner with great fecundity, that feeds on benthic fauna, often preferring molluscs. It may reach a length of 25 and occasionally over 30 cm.

Population in Europe

Generally populations are more stable in eastern than in northwestern Europe. Backwaters overgrown with submerged plants are a prerequisite for a stable population.

Reasons for decline

With the drainage of swamps and the pollution of pools and backwaters adjacent to the main streams, this species has disappeared completely from large areas in Central Europe.

Conservation measures taken

No conservation measures habe been taken. However, this species has been indirectly helped by bird conservation and particularly conservation of habitats suitable for herons (*Nycticorax nycticorax* (L.), *Botaurus stellatus* (L.) and *Ixobrychus minutus* (L.)) where the weatherfish commonly occurs, as well. It is occasionally protected by law, for example, in The Netherlands.

Conservation measures proposed

Protection and conservation can be easily achieved by habitat conservation. It would be sufficient to protect individual localities with good stocks of this fish. Re-introduction into localities where it formerly occurred is strongly recommended.

Noemacheilus barbatulus (Linnaeus, 1758)

E. Stone loach
F. Loche franche
G. Schmerle

Vernacular name: –

Family: Cobitidae

Classification: rare to vulnerable

Note: Four subspecies are described in addition to the nominate species: *Noemacheilus b. caucasicus* (Berg, 1899) found in the Argun River, a tributary of the Terek, *Noemacheilus b. sturanyi* (Steindachner, 1892) in Lake Ohrid, *Noemacheilus b. vardarensis* (Karaman, 1928) in the Vardar Basin and *Noemacheilus b. hispanica* (Bacescu-Mester, 1967).

Distinguishing characteristics

The stone loach is an elongated fish, with a body covered with very small, deeply embodied scales, only moderately laterally compressed. The mouth is inferior and has six barbels, four under the tip of the snout and one at each corner of the mouth. The fins are rounded, the insertion of the dorsal fin is slightly in front of or above the base of the pelvic fin. The colouring is extremely variable. The dorsum is dark (usually dark-brown), the sides are lighter and the belly creamy yellow. Individuals from clean waters often have irregular marble patterns on the sides. The colouring is less developed in places where the water quality is poor and in large streams where the fish live in deep water. Meristic data: D 10 to 11, A 8 to 9.

Distribution in Europe

The species occurs throughout Europe, from the east of the Pyrenees, Oliver (1970), except for the Iberian Peninsula, central and southern Italy, Greece, and Scotland. It also occurs from southern Sweden and Finland, east to Siberia and the Volga, Ural and Emba Basins; for details see distribution map, p. 257.

Ecology

The stone loach inhabits more or less the same river stretches as the brown trout and grayling, but is also occurs in the lower reaches. As a rule, it first appears downstream from the mountain torrents occupied exclusively by trout and bullhead. It is most common in the central trout zone. Occasion-

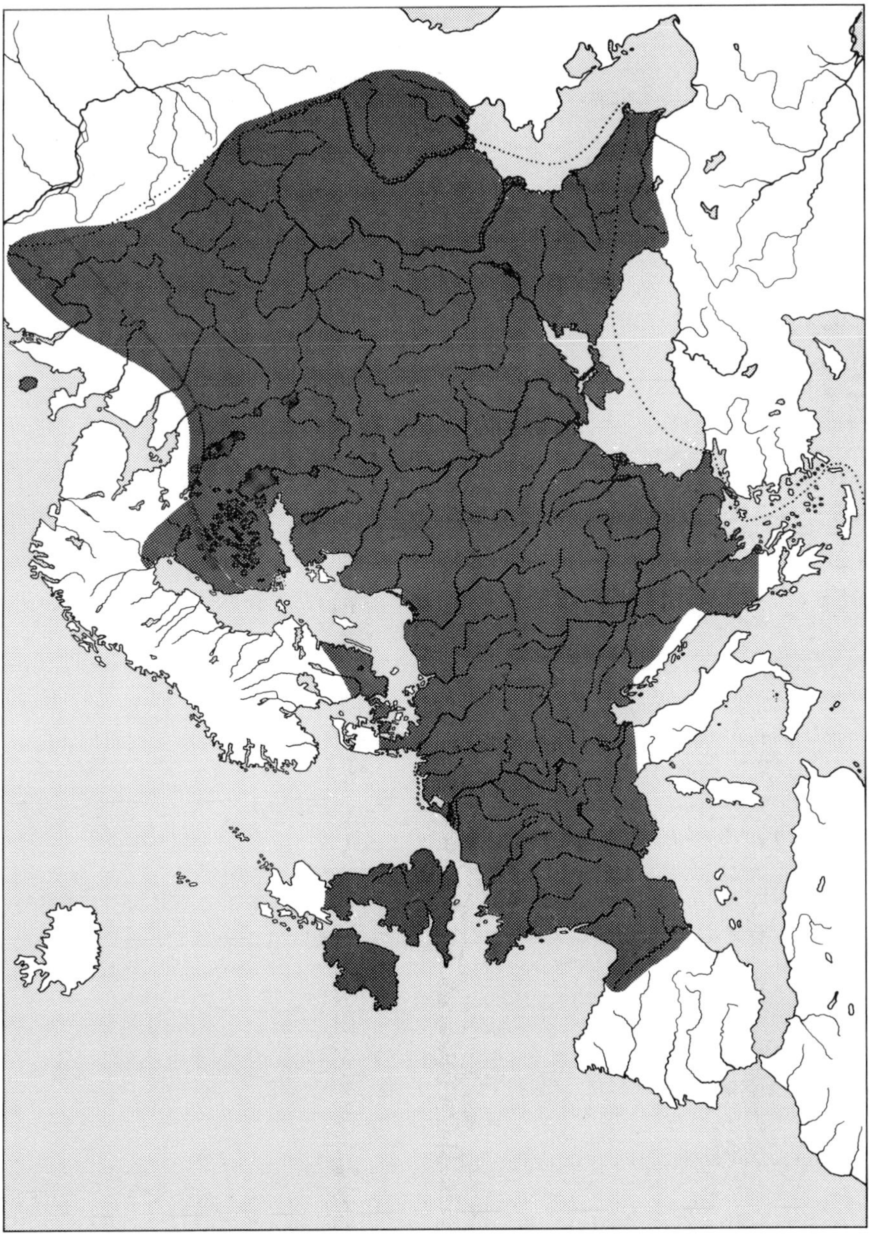

Fig. 66: Distribution of *Noemacheilus barbatulus* (Linnaeus, 1758).

ally it may live in lowland stretches. Its habitats are always lotic sections, preferably with diversified bottom and little or moderate silting. This bottom dwelling species, exhibiting only local movements of about 500 meters, are regularly dispersed in shallow waters in places, where the substrate is suitable. The mobile fraction of the population does not exceed 25%. Exceptionally, the movement of one individual fish reached almost 10 kilometers from the place of its marking (BRUNKEN 1987). Spawning starts in April and sometimes lasts for two months. Where sandy substrates, suitable for spawning occur a great many mature individuals may be present, though they never congregate in shools. The eggs are usually released in successive batches, generally three. Growth is rapid during the first year of life and individuals may reach 7 to 8 cm. Maturity is reached in the second year of life. The life-span is short, maximum of five to six years. The diet is typical of bottom dwellers and includes mostly invertebrates and detritus. If this species occurs in trout streams, as it did regularly about 50 years ago, it could help to restore balanced fish populations in the Central European submontane regions. It is important as a food fish for larger brown trout and itself may control some cyprinids, eg. the chub and the dace by eating their spawn and freshly hatched larvae. For further information see SMYLY (1955) and LIBOSVÁRSKY (1957, 1959) and BRUNKEN (1987). Richly oxygenated water is a necessity, but the fish can tolerate a rather high degree of organic pollution, except in lowland areas, such as The Netherlands (Dr. LEENTVAAR pers. comm.). Industrial pollution and detergents have eliminated this species in long stretches of Central European rivers.

Population in Europe: Locally abundant or locally absent.

Reasons for decline

Pollution is more responsible for the decline than changes to the environment.

Conservation measures taken

No conservation measures known, except legal protection in some European Countries.

Conservation measures proposed

Pollution control could help dramatically, and re-introduction causes no difficulties.

3.14. Siluridae

Silurus glanis (Linnaeus, 1758)

E. Wels
F. Silure glane
G. Wels, Waller

Vernacular name: –

Family: Siluridae

Classification: rare to vulnerable

Distinguishing characteristics

The wels does not need to be described in detail, as its broad, flat head with six barbels, two of which originate on the upper jaw, four on the lower jaw, render it unmistakable. The dorsal fin is very short, the anal very long and joining the caudal. It has no adipose fin. The colouring is very variable. The dorsum is always dark (from olive-brown to blue-black). The sides are lighter, often with marble patterns, the underside is greyish-white, often with dark moiré. Meristic data: D 3 to 5, A 72 to 92. The body is scaleless.

Distribution in Europe

The European range includes the Danube Basin, all rivers emptying into the Baltic Sea, and the Elbe River. Further the wels is indigenous to Lake Constance and the Murtensee. Before attempts to extend its distribution were made, it was observed in the Upper Rhine near Breisach in 1858 and downstream in the vicinity of Strasbourg in the 16th century (SIEBOLD 1863). LAUTERBORN (1916) mentioned its presence in the Upper Rhine and its rare appearance in the Doubs River, and explained this as result of past changes in geomorphology and glaciation (see also ZSCHOKKE and STEINMANN (1911). To the east, the wels occurs in rivers feeding the Caspian, Azov and Aral Seas. It is rare in Jutland, southern Sweden and Finland. It is found in the Neva river system and Lake Ladoga. The southeastern range limit is the Vardar River Basin. It was originally absent from England, though introduced less than 100 years ago (VARLEY 1967). It originally does not occur on the Iberian Peninsula, in France (except in the Doubs), Italy, and rivers flowing into the Arctic. See distribution map, p. 260.

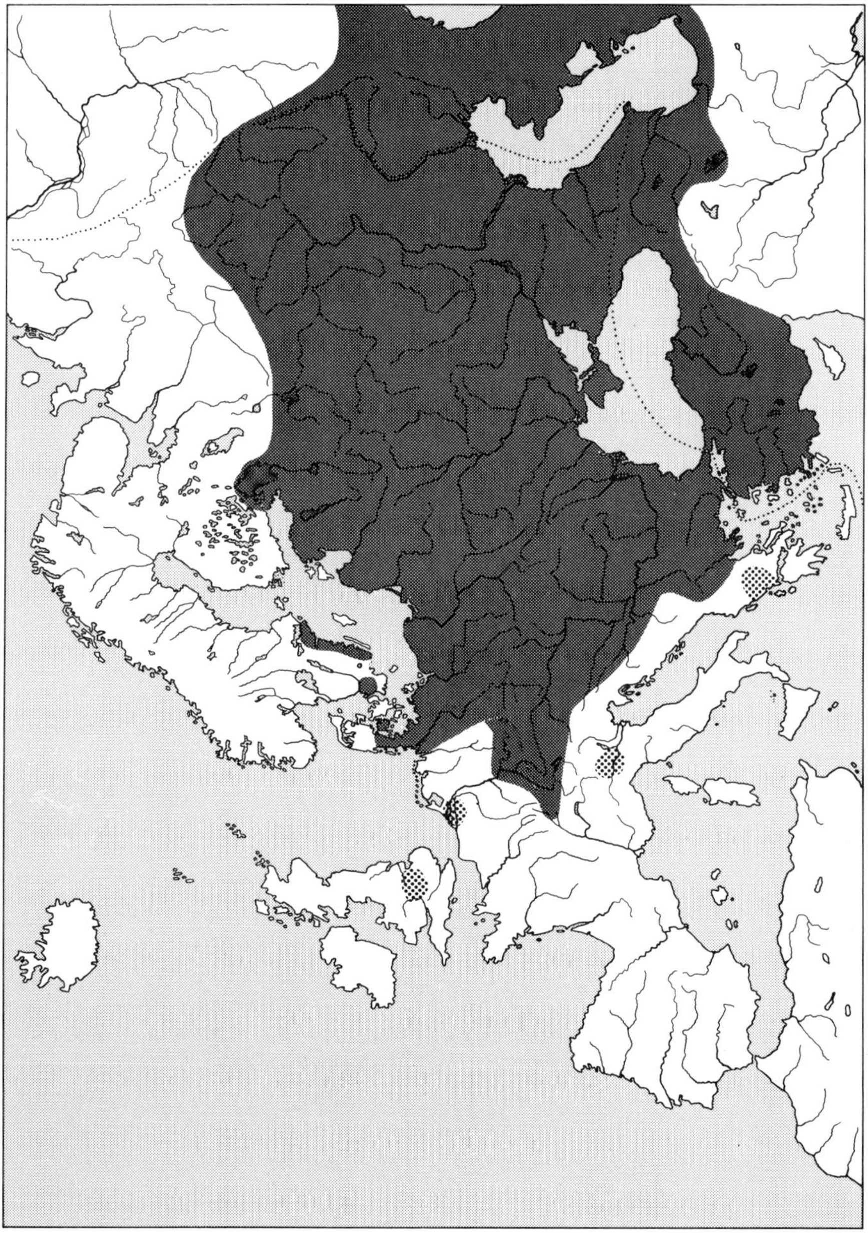

Fig. 67: European distribution of *Silurus glanis* (Linnaeus, 1758).

Ecology

The wels is one of the largest fish of European waters. It therefore requires large rivers, lakes, backwaters and lagoons. It can tolerate brackish water, for example in the Baltic Sea. Adults live alone, resting in holes in the river bed beneath overhanging banks, sunken trees, etc. It is most active at twilight and throughout the night. During the spawning period, it travels short distances to find suitable places to spawn. It prefers tree roots in medium depths. During floods, it spawns on grass, provided the water temperature reaches at least 18° C. The female builds a primitive nest from plants. The eggs are attached to them, and the male guards the spawn for some time. The releasing mechanism for spawning activity, provided the substrate, temperature and environment are suitable, is a sudden drop in barometric pressure, as when a warm and sticky day is followed by a thunderstorm. Knowing this, it is relatively easy to induce the wels to spawn semi-artifically for stocking purposes. The young soon take benthic invertebrates, and shortly after they start to feed exclusively on fish. Large individuals may consume other vertebrates as well. Growth may be fast, but depends on the food supply. The largest individuals may reach 3 to 4 m. In the Danube Basin, fish over 1 m and weighing 30 to 60 kg were no exception. After spawning, the adult fishes move back to deep water. Winter is spent in deep holes and crevices in the river bed. In the lakes, the winter hibernation places are never in the deepest spots. Individuals remain in the lower third of the water column, on soft mud (Lelek et al. 1964). Requirements for oxygen concentration are not high, and it is not exceedingly sensitive to pollution.

Population in Europe

In general, its occurrence is sporadic at the western and northern edges of its range. It is extremely scarce in Scandinavia; some populations are on the brink of extinction. In the Danube Basin, except where the river has been modified for navigation in its upper reaches, there are stable populations. Because of heavy fishing (particularly angling) it is dependent on stocking.

Reasons for decline

Man-made changes along rivers, particularly straightening and reinforcement of banks, endangered the natural reproduction; in some small streams heavy pollution also has had an adverse effect.

Conservation measures taken

Legal size limits and closed seasons have been introduced in almost all countries where the wels occurs. The creation of sanctuaries, particularly for wildfowl, and protected spawning areas proved to be very useful to promote natural reproduction.

Conservation measures proposed

The first step to be taken to increase the populations within its natural distribution area is stocking with pond-reared juveniles. Secondly, protection of all remaining natural and/or semi-natural habitats is necessary. The legal size-limit must be set at no less than 90 cm, assuming that maturity is reached at the age of 5 to 6 years by fish ranging from 50 to 80 cm in length. Larger populations of wels could help to restore balanced fish populations. At present, our waters are partly crowded with little wanted cyprinid species, such as roach, white bream and bream which could be limited by a predator like the wels.

3.15. Gadidae

Lota lota (Linnaeus, 1758)

E. Burbot
F. Lote de rivière
G. Quappe, Rutte

Vernacular name: –

Family: Gadidae

Classification: rare to intermediate

Note: This species is represented by a subspecies *Lota l. maculosa* (Le Seur, 1817) in northeast Siberia. On the American continent, recognition of this subspecies seems unwarranted.

Distinguishing characteristics

The burbot has all the typical gadid characteristics: all rays soft, minute scales and one barbel on the chin. It has two dorsal fins. The barbel on the lower jaw is long. A raised rim on the anterior nostril appears to form a short barbel. The jaw extends as far as the hind edge of the eye. The pelvic fins are located in front of the pectorals. The colour is extremely variable according to the habitat and season. The dorsum is usually greenish-brown, with grey mottling. The belly is light. Meristic data: D1 10 to 15, D2 71 to 88, 59 to 63 vertebrae.

Worldwide distribution

The burbot is found in all suitable freshwater habitats of Eurasia and North America, southward to about 40° N. It is absent from Kamchatka. See distribution map, p. 264.

Distribution in Europe

The burbot is found almost everywhere in Europe north of the Balkan Peninsula and the Pyrenees. It is found in the Rhône, but is very rare in the Seine and the Loire. In England, it probably occurs only in the eastern part of the country and seems to be nearing extinction (MAITLAND 1972, WHEELER 1969). In Italy, it occurs in the Po Basin and Lake Garda, and is

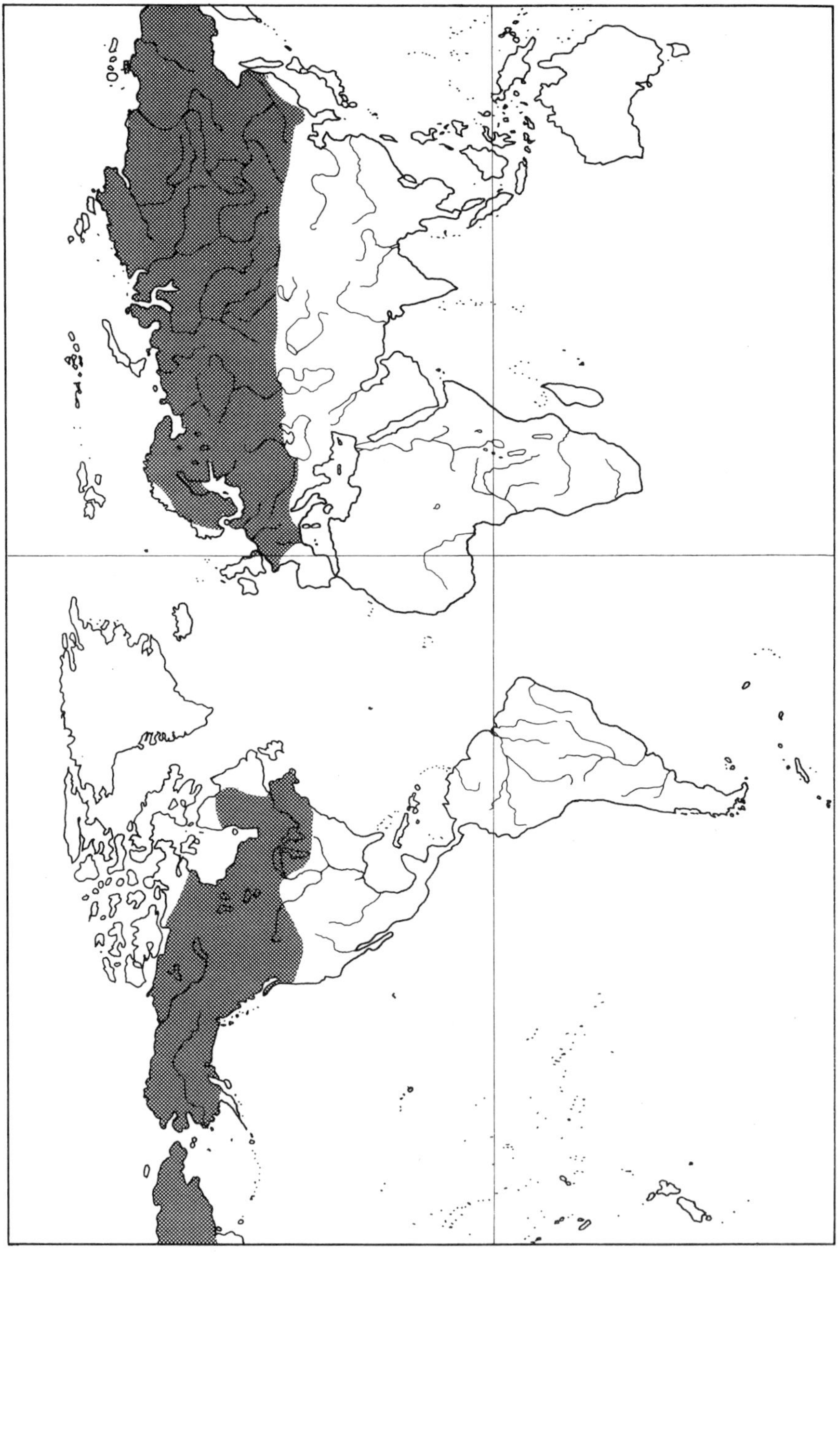

Fig. 68: Worldwide distribution of *Lota lota* (Linnaeus, 1758).

present in the whole Danube system. Only in the most western and northeastern Norway, Iceland and the Arctic islands it is entirely absent.

Ecology

This fish generally lives on the bottom of clear rivers and lakes, under rocks or other objects, or in holes of the banks. Both, juveniles and adults are nocturnal. In Central Europe, breeding takes place in winter, from December to February. Fecundity is very high. It is a litho-pelagophilic spawner. The eggs are semi-buoyant due to a large oil globule in the yolk. Slight water movement can lift or move them over the bottom. The burbot matures in its third to fourth year. The young fish live chiefly on small crayfish, *Asellus,* and the larvae of most insects. Large burbot prey on fish, including perch, gudgeon and stone loach, preferring bottom dwelling species. This species has successfully colonized several cool and clean man-made lakes. During the summer, it falls into semi-lethargy. It can only be seen in winter, and particularly at spawning time, it loses its alertness. During this time, individuals congregate under transparent ice on a clean, sandy bottom and can easily be caught there. Because of its nocturnal habits, its distribution is probably inadequately known. In addition, its supposedly high oxygen requirements are perhaps exaggerated. Recently (1978) several specimens were observed in old branches of the River Rhine at km 475.

Population in Europe

Records on the size of populations are available only from northern and eastern Europe, where this species is commercially exploited. In general, little is known about this species in other parts of its distribution area.

Reasons for decline

Decline has been reported only from England, where only a few specimens have been seen in recent years (Maitland, pers. comm.). Lack of information, especially older information, prevents any comparison with the present situation.

Conservation measures taken

In some countries a size limit exists.

Conservation measures proposed

The authorities involved in managing salmon streams must be convinced that the burbot is not the worst predator. The high losses of freshly released salmonids caused by burbot could be avoided by conditioning the hatchery-reared trout and salmon. Furthermore, fishes should not be released free immediately after being transported from elsewhere, as at that time they are exhausted and easily fall prey to burbot. Reduction of artificially high maintained stock of the eel would help to increase the abundance of the burbot.

3.16. Gasterosteidae

Gasterosteus aculeatus (Linnaeus, 1758)

E. Stickleback
F. Epinoche
G. Dreistachliger Stichling

Vernacular name: –

Family: Gasterosteidae

Classification: intermediate to rare, locally vulnerable

Note: Three or four basic forms, Gąsowska (1962) mentioned 8 forms, are described. The definitive characteristic is the number and position of bony scutes on the sides. The form *trachurus* bears the scutes along the whole lateral line, while *semiarmatus* has the scutes only on the sides and caudal penduncle; *leiurus* has the scutes only on the anterior section of the sides; *hologymnurus* has only one scute in the middle of the body.

Distinguishing characteristics

The stickleback has a rather elongated body with a particularly long head and snout. There are two dorsal fins. The first is composed of three isolated spines, the second dorsal and the anal fins are opposite each other and similar in shape. The pelvic fins are reduced, but have one long, sharp spine. Colouring varies from silvery green to almost black on the dorsum, shading through silver on the sides to white on the belly. In the spring, males have brilliant red colouring of the isthmus and belly; the dorsal part is green; the eyes clear blue. Meristic data: DIII, 8 to 14; A I, 6 to 11.

Worldwide distribution

The stickleback is nearly circumpolar but absent from the Arctic Coast of Siberia, most of the Arctic Coast of North America and it is widely distributed in the northern hemisphere in both salt and freshwater. It is either anadromous or a purely freshwater species. In Europe, it is found in southern Greenland, Nowaya Zemlya, northern Norway and Iceland, along the coast southward to Spain, and in the Mediterranean and Black Seas. It is absent from the Basins of the Caspian and Aral Seas.

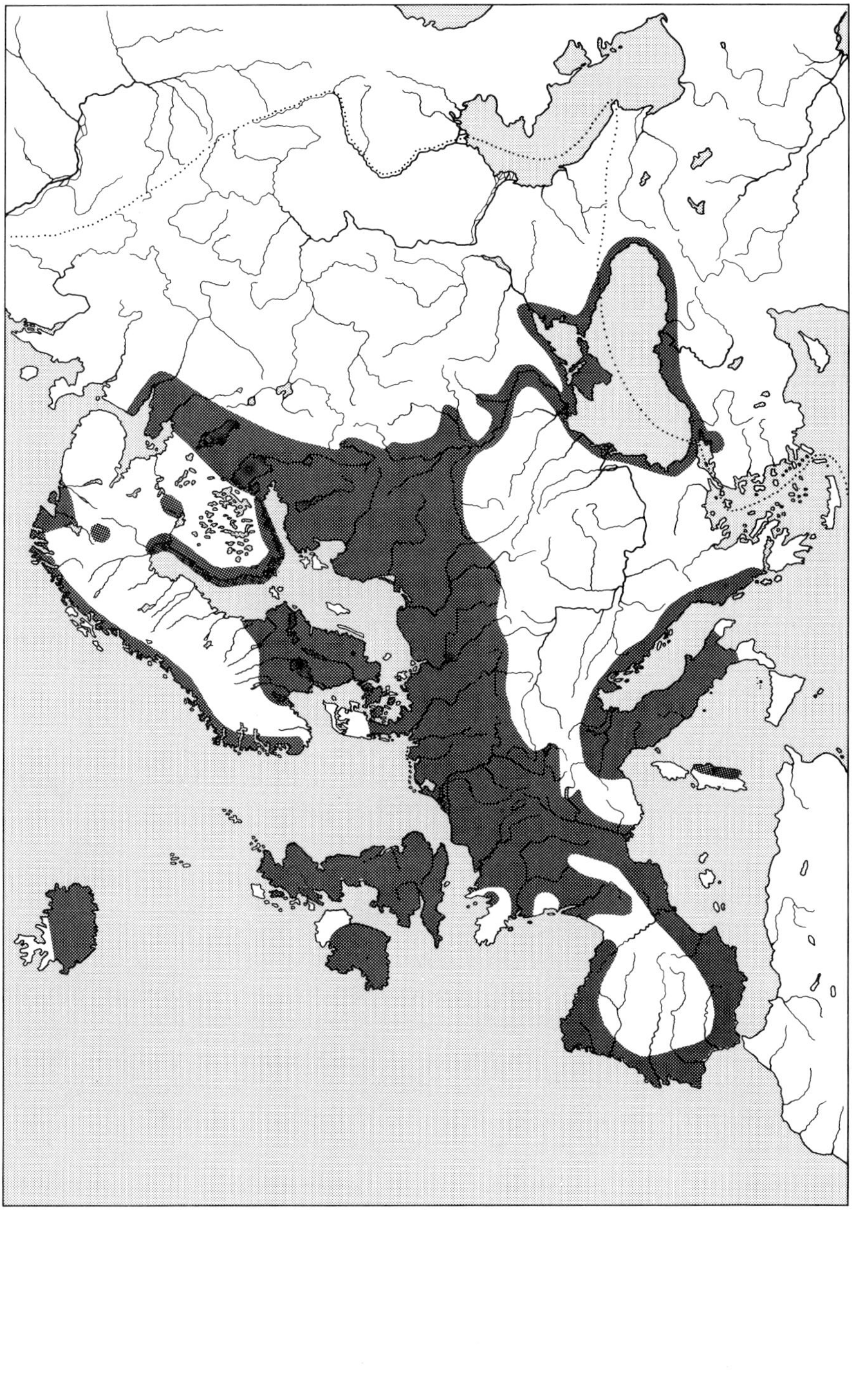

Fig. 69: European distribution of *Gasterosteus aculeatus* (Linnaeus, 1758).

Distribution in Europe

It is absent from central Scandinavia, central areas of the Iberian Peninsula, and the Danube Basin. Solitary individuals caught there are very probably of foreign origin as was the case in the Upper Danube, downstream from Vienna, and probably on Sardinia (Bianco 1980). See distribution map, p. 268.

Population in Europe

The population in Europe is moderate to large in coastal areas where there is little pollution. Inland, the freshwater distribution is not continuous, and it is often abundant at isolated localities.

Ecology

In northern Europe, including some coastal areas, the stickleback is probably the most common species of freshwater fish. It was found in virtually all waters except fast flowing streams. The reproductive biology has been frequently described. The male builds a nest, composed of plant material, stuck together with a kidney secretion, in a shallow depression on the bottom, usually on sandy substrate. The female is driven into it to lay the eggs. The male guards the spawn, fans the eggs with its pectoral fins and removes those, that are dead or infertile. Growth is fast; at the end of the first year, it reaches 2 to 5 cm in length. Maturity is reached after the first year of life, and the fish rarely lives more than 3 to 3½ years. Freshwater populations grow more slowly. The main spawning season is April and May but it often continues throughout the summer. The food varies according to what is available. It generally takes any prey of suitable size. Vegetable food is probably taken by accident.

Reasons for decline

Inland populations have without any doubt suffered from pollution and increased silting at suitable spawning places. There are large areas in Central Europe, such as the Rhine and its tributaries, where this species has become locally rare.

Conservation measures taken

No conservation measures have been taken.

Conservation measures proposed

It is thought that more stickleback would be useful to redress the ecological balance, particularly among the present 3 to 4 species fish populations of polluted Central European streams. A greater abundance of this fish may promote the development of more balanced fish populations. Reduction of pollution and restoration of weeded lentic habitats with patches of clean bottom along navigable rivers might help to increase populations within inland waters.

Pungitius pungitius (Linnaeus, 1758)

E. Ten-spined stickleback
F. Epinochette
G. Zwergstichling, Neunstachliger Stichling

Vernacular name: –

Family: Gasterosteidae

Classification: intermediate to rare, locally vulnerable

Note: One subspecies *Pungitius p. hellenicus* Stephanidis, 1971 is known from the Sparchilos River, Greece.

Distinguishing characteristics

This species bears some basic similarities to the common stickleback, *Gasterosteus aculeatus*. Its most obvious characteristic is the number of spines in front of (or in) the first dorsal fin: there is an alternately spaced row of eight, nine or ten isolated small spines, all smaller than the eye diameter. The pelvic fin has a single large spine. The colour is variable, the dorsum and sides are usually olive-green, occasionally with dark bars on the sides. It is silvery underneath. Meristic data: D_1 VII to XII, D_2 9 to 12; 32 to 33 scales in lateral line.

Worldwide distribution

P. pungitius is generally found in fresh and salt waters throughout the northern Hemisphere. In the coastal areas, it usually stays close to the shore. In North America, it ranges from the New Jersey Atlantic Coast, north along the entire Atlantic and Arctic Coasts, westward to the Aleutian Islands and along the Pacific coast to the Kenai Peninsula (Cook Bay), but no further south. Inland it ranges from the southern Great Lakes Basin northwards to the Arctic Coast.

Distribution in Europe

The ten-spined stickleback is regularly found along the coasts of northern Europe and occures scattered deep inland, sometime 200 to 300 km off the coast including the Baltic drainage. It is also found in inland waters throughout Ireland, Scotland and England, including the Hebrides. It is absent from Iceland, the Balkan Peninsula, the Alps, the Pyrenees and

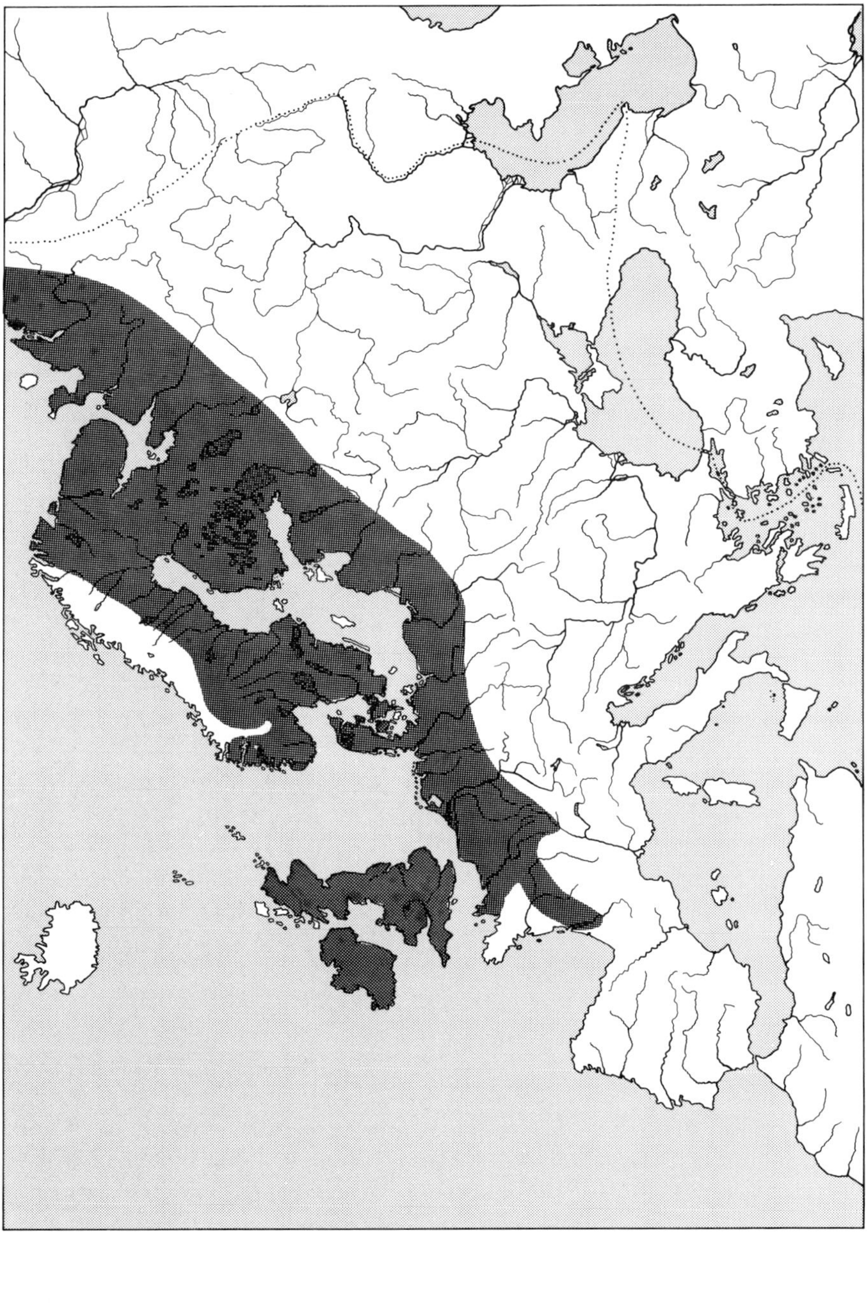

Fig. 70: Distribution in Europe of *Pungitius pungitius* (Linnaeus, 1758).

much of the territory north of those mountains. Little is known of the eastern distribution limits. A tentative distribution map is provided on page 272.

Population in Europe

In general this species is less abundant than the common stickleback and does not occur so far inland. It has never been recorded from higher altitudes.

Ecology

Its ecology is generally similar to that of *Gasterosteus aculeatus*. A nest is built for reproduction but not necessarily on the bottom but more often in the waterweeds. This species prefers dense weeds and quiet water. It lives mainly in small water bodies, such as pools and ditches. The lifespan is shorter than that of the common stickleback, about three years at maximum.

Reasons for decline

The same as for the common stickleback, in addition, because weedy and swampy areas near the estuaries of large rivers, and flood prevention installations have been drained.

Conservation measures taken

No conservation measures known.

Conservation measures proposed

Conservation measures would be difficult to realize as the extension of flooded areas in northern Europe seems unrealistic. It is hoped that this species will survive in areas near to the dams, in brackish waters and in a few localities near estuaries. Inland, some protected damp regions may be refuge for isolated populations.

3.17. Cottidae

Cottus gobio (Linnaeus, 1758)

E. Bullhead
F. Chabot
G. Groppe

Vernacular name: –

Family: Cottidae

Classification: vulnerable

Note: In its easternmost distribution area, in the Dnieper, Neva, Volga and Ural Rivers, the bullhead is represented by the subspecies *Cottus g. koshewnikowi* (Gratzianov, 1907). *Cottus g. petiti* was described from the Pyrenees, and *Cottus g. ferruginosus* found from Lake Garda to Dalmatia, is sometimes considered as a species.

Distinguishing characteristics

The bullhead has a flattened head and a single spine on each gill cover. The body is scaleless. Both pectoral and pelvic fins are located closely to the operculum. The pelvic fins are light in colour, not barred. The inner ray is equal to or slightly shorter than the outer ray. The pelvic fins may reach to the vent, but are usually shorter. The colour is variable, sometimes heavily mottled, sometimes with distinct bars running across the sides. All the fins except the pelvic are dark with irregular light lines. Meristic data: D_1 6 to 9, D_2 15 to 20, A 10 to 14.

Distribution in Europe

This species occurs from northern Spain throughout the rest of Europe. It occurs in England, parts of Sweden and Finland, eastward to the Petchora Basin, and in coastal waters of the Baltic Sea. The range extends southwards to the Alps, the lakes of the Rhine, the Rhône and Po Basins, and the Attersee. In Italy, it is found in the Tiber Basin but not in Sicily. It occurs in Dalmatia, but is absent from the Balkan Peninsula, except possibly for the Vardar River. It is also absent from Norway, northern Sweden, most of Finland, Iceland and the Kola Peninsula. See distribution map, p. 275.

On the continent of North America, the bullhead is replaced by related species including *Cottus ricei* (Nelson 1976) and several others.

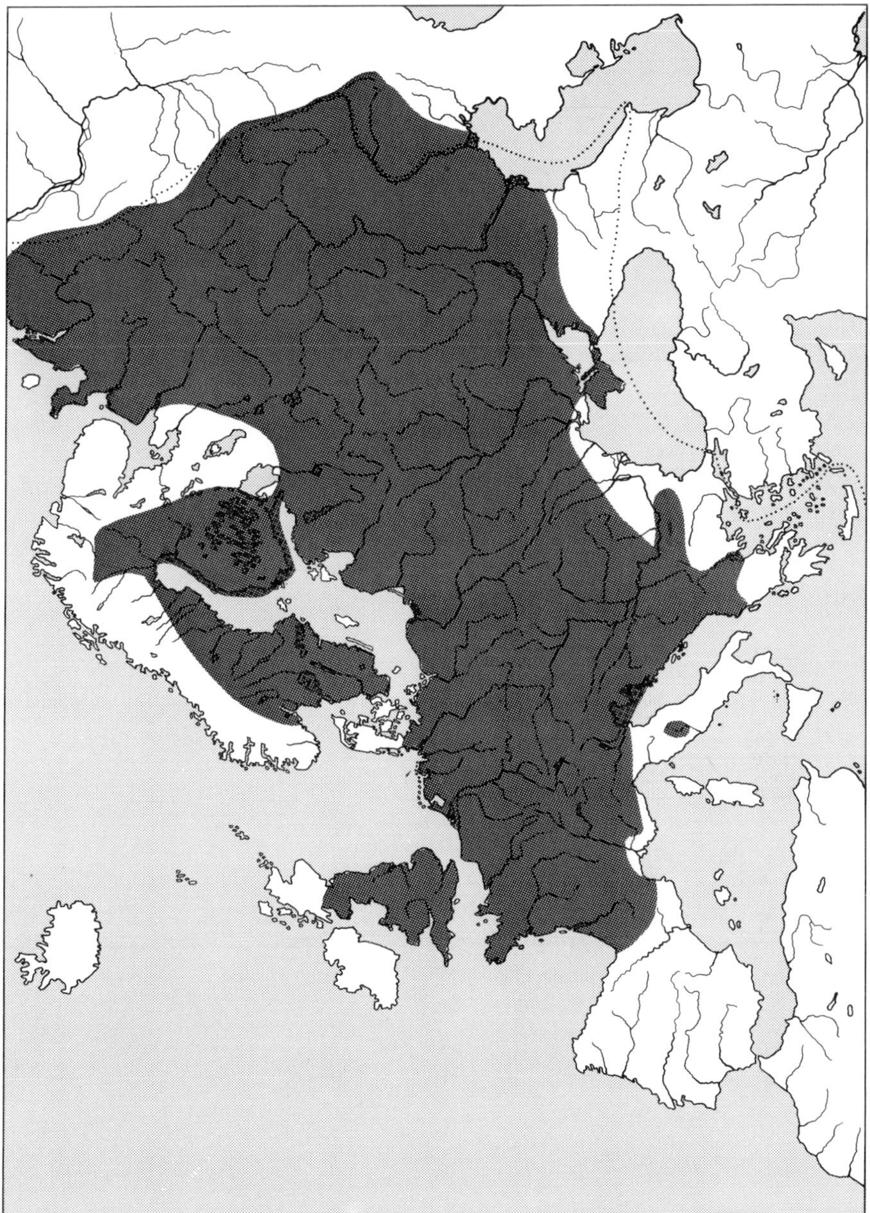

Fig. 71: Distribution of *Cottus gobio* (Linnaeus, 1758).

Ecology

The bullhead is common in unpolluted streams, rivers and occasionally lakes in which the bottom is stony, the water temperature low, and the water always saturated with oxygen. The bullhead is the most typical species of the brown trout zone, occurring in tiny torrents reaching altitudes over 2000 m. It prefers shallow water, where it lives between the stones and other objects on the bottom. The lifespan is short, 5 years at maximum. Sexual maturity is reached in the second year of life. Spawning takes place in shelters under the stones. The eggs are attached to the walls of a shelter, and the male guards the eggs until they hatch. It feeds on benthic invertebrate fauna. Until recently, the bullhead was considered a competitor of the brown trout for food, but detailed stomach analyses, in which food taken was identified down to species level, indicated that competition is almost negligible (Straškraba et. al 1966). This fish reaches about 3 to 4 cm in length in the first year of life and a maximum size from 9 to 10 cm. This stationary species may be considered as an indicator of good quality water.

Population in Europe

The bullhead formerly occurred in almost all waters inhabited by brown trout. Its present population is influenced by the management of trout streams because it is often removed as a competitor of the trout. It has disappeared from many river stretches because of pollution.

Reasons for decline

This is one of the species most sensitive to physical and chemical changes of water, which is why it has disappeared from almost all the lower reaches of large trout streams, as these have become polluted. Modifications, particularly changes to the river bed producing uniformity on the stream bottom, are also harmful. Management of trout streams as reasons for decline has already been mentioned.

Conservation measures taken

No active conservation measures are known. Only in some Scandinavian and in several European countries this species is protected by law.

Conservation measures proposed

This species may be successfully and permanently protected only in river sanctuaries where there is no water pollution and any yield oriented trout management. Detailed information concerning the necessary bottom substrate and water quality requirements is provided by Bless (1983).

Cottus poecilopus (Heckel, 1836)

E. Siberian sculpin
F. –
G. Sibirische Groppe

Vernacular name: –

Family: Cottidae

Classification: rare and vulnerable

Distinguishing characteristics

All characteristics of the bullhead are possessed by this species as well, but it should be noted that the pelvic fins are always barred and the inner ray of the pelvic fin never reaches half the length of the outer one, and usually does not exceed 1/3 of its lenght. The pelvic fins reach the anal opening and often extend beyond it. Meristic data: D_1 8 to 9, D_2 17 to 19.

Distribution in Europe

Dependable information on the distribution of the Siberian sculpin are available, particularly from Sweden (Müller 1960, Andreasson 1971). The data presented by Duncker and Ladiges (1960) need to be up-dated for northern Europe. This sculpin is also found in parts of the Oder and Vistula Basins and in some Carpathian streams, leading to the Danube, Prut and Dniester. A tentative distributiom map is provided on page 279.

Ecology

Similar to that of the bullhead. Moreover, wherever both species occur together, *Cottus poecilopus* occupies the stretches located upstream from those of the bullhead. It requires cool, well oxygenated waters. It has been found that *Cottus poecilopus* cannot permanently live in waters with less than 8 mg O_2/l.

Conservation measures

See bullhead.

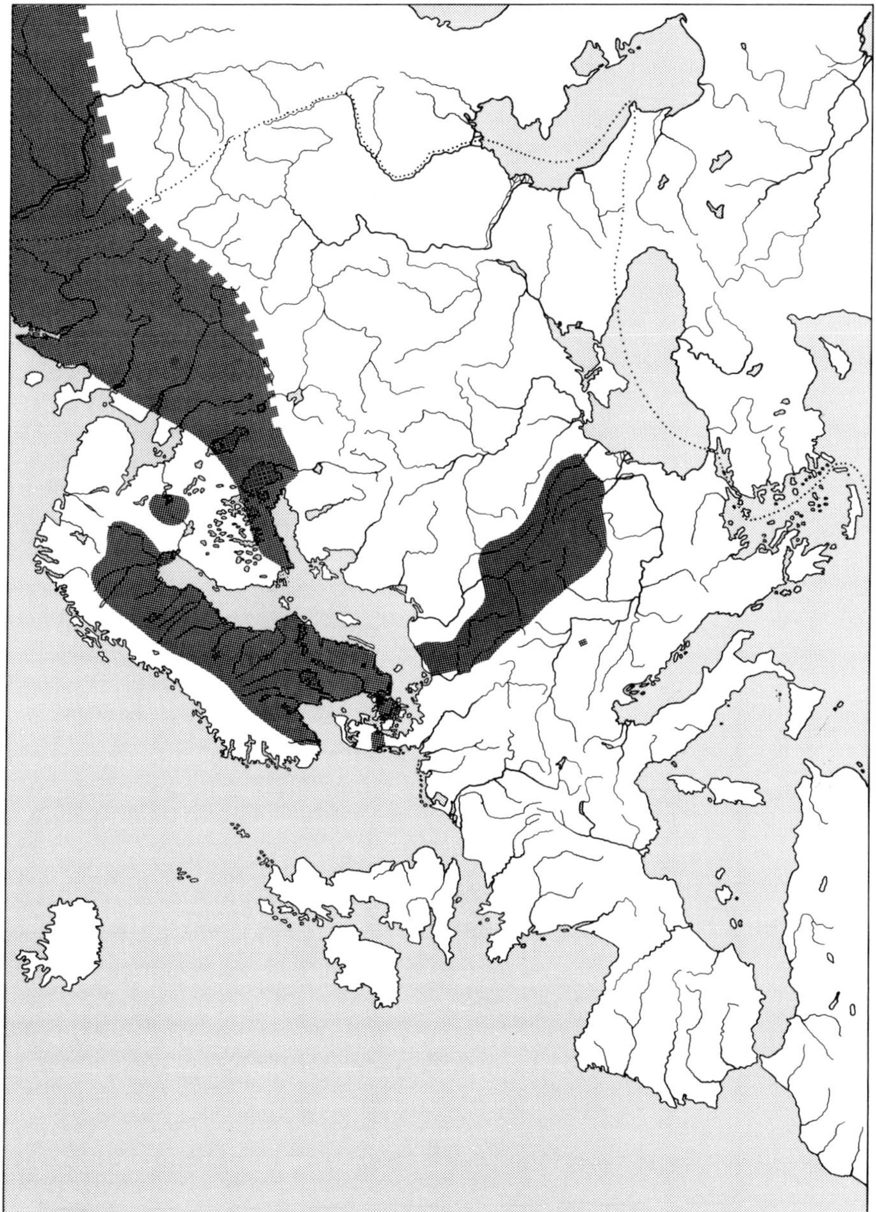

Fig. 72: Distribution of *Cottus poecilopus* (Heckel, 1836) in Europe. Jagged line in eastern Europe: range of the siberian sculpin not clear.

3.18. Percidae

Gymnocephalus (Acerina) acerina (Güldenstädt, 1975)

E. –

Vernacular name: Donskoi ersh (Russian)

F. –

G. Don Kaulbarsch

Family: Percidae

Classification: vulnerable

Distinguishing characteristics

This typical percid fish of the genus *Gymnocephalus* has some morphological similarities to *Gymnocephalus baloni* or *G. schraetser,* particularly regarding the general body shape. Meristic data: D_1 17 to 19, D_2 12 to 14; A 7 to 8; 50 to 55 scales in the lateral line. For detailed information see Oliva (1959) and Collette and Bănărescu (1977).

Distribution

This species is confined to some rivers flowing to the Black Sea from the North, particularly the Rivers Dniester, Don, Southern Bug, Dnieper, Donets, and Kuban. See distribution map, p. 281.

Ecology

Similar to that of *Gymnocephalus schraetser* of the Danube catchment area. It lives in swiftly flowing water with sandy or gravelly bottom.

Conservation measures taken

Not known.

Conservation measures proposed

Protection of habitat within selected stretches with well established populations.

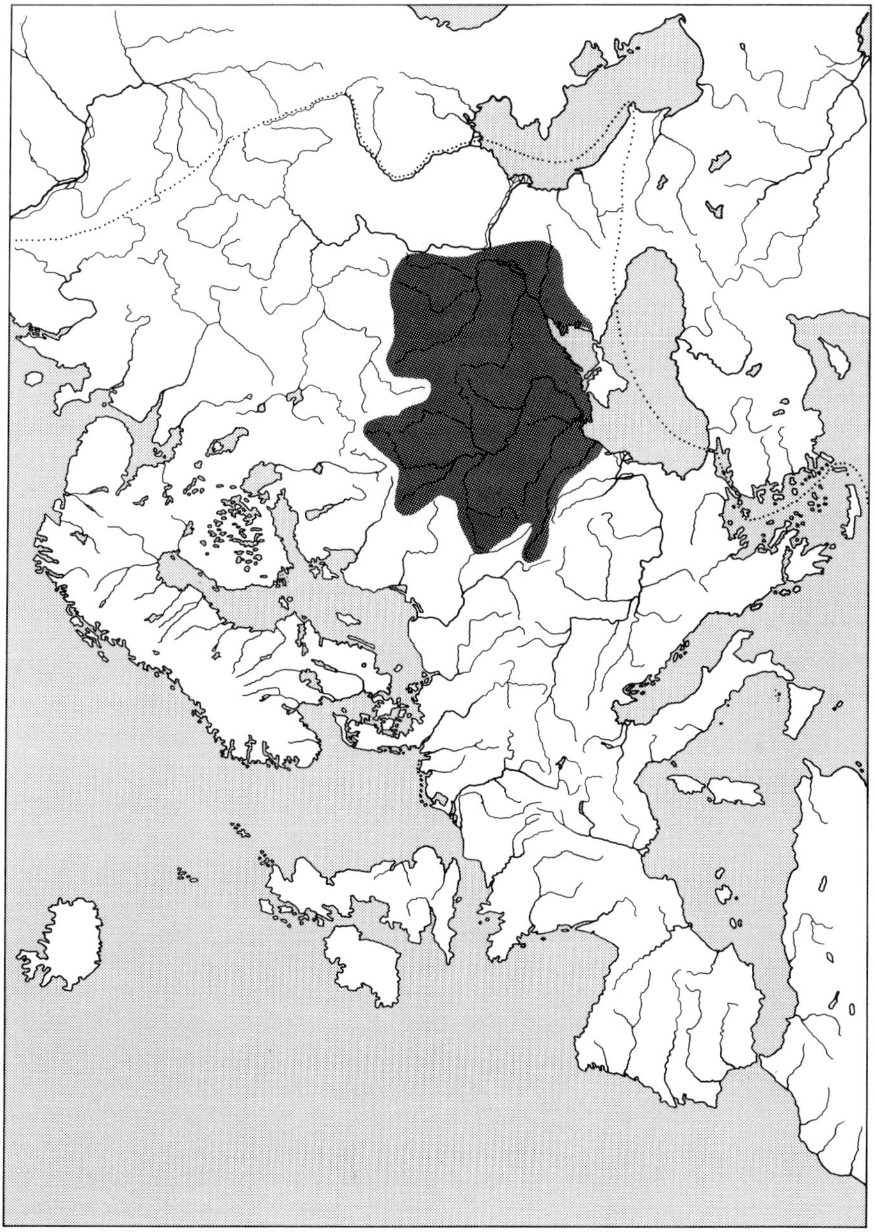

Fig. 73: Distribution area of *Gymnocephalus acerina* (Güldenstädt, 1975).

Gymnocephalus baloni Holčík and Hensel, 1974

E. Balon's ruffe

F. –

G. –

Vernacular name:
Hrebenačka Balonova (Slovak)

Family: Percidae

Classification: vulnerable

Distinguishing characteristics

Balon's ruffe is a typical small percid fish like the common ruffe *(Gymnocephalus cernuus)*. It is difficult to distinguish this species from the common ruffe or other species of this genus. It has four to six dark transverse bands on the sides, sometimes vaguely delimited. The body is robust, short and deep. The nape covered by scales, the opercule has two spiny protuberances. Meristic data: D_1 XIV to XV (XVI) D_2 11 to 12; pored scales in lateral line: (35) 36 to 37 (39); gill rakers: (9) 11 to 13 (14), short; preopercular spines: (9) 10 to 13 (14).

Ecology

Detailed information is not available, though some preliminary observations indicate that it is exclusively rheophilic. The holotype and paratype of this species were caught in the Danube among stones washed by the strong current. The eroded fins observed on most specimens give evidence of its rheophilic habits.

Distribution

According to preliminary investigations, this species seems to be limited to the Danube Basin. It has been found in the Danube in Czechoslovakia and in its delta in Romania, in the Bavarian Danube above Vilshofen. *Gymnocephalus baloni* obviously inhabits the lower stretches of some large rivers in the Danube Basin. Its presence in the Latorica (Slovakia) and Timisal River in Romania has been recorded. It is presumed that it is an endemic Danubian species, being found mostly in the middle and lower Danube, and in the middle and lower course of some of its tributaries, but the exact range is not yet known.

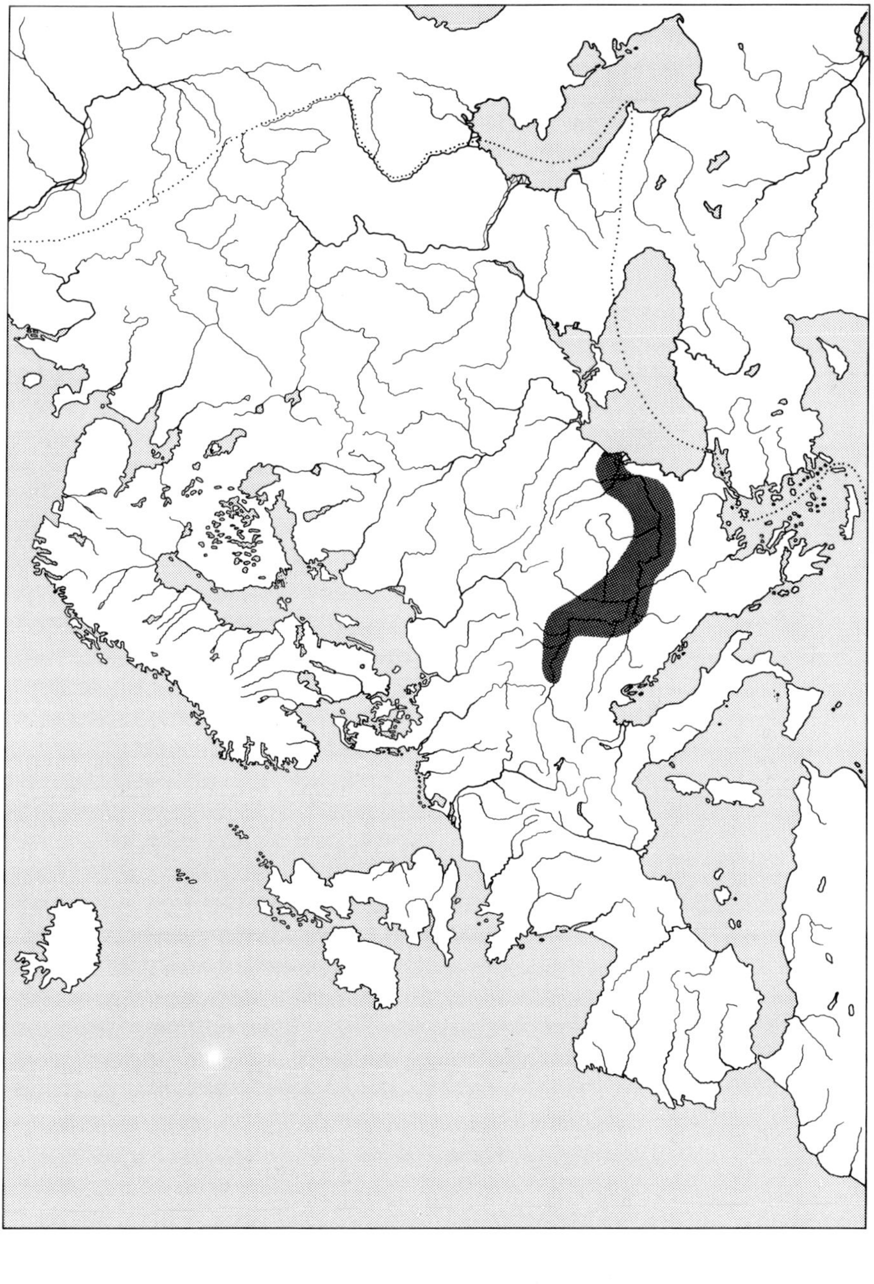

Fig. 74: Distribution of *Gymnocephalus baloni* (Holčík and Hensel, 1974).

Population in Europe

The only more detailed informations on its populations in Europe are available in Holčík and Hensel (1974).
It is suggested that this species might also be found in the Dnieper River in the USSR, outside the known distribution area within the Danube System.

Reasons for decline and conservation measures

No information on its possible decline or desirable conservation measures is available. It is assumed that this species is sensitive to man-made changes, particularly to modifications of the river course and damming, as is the case with all rheophilous species.

Gymnocephalus (Acerina) cernuus (Linnaeus, 1758)

E. Ruffe
F. Grémille
G. Kaulbarsch

Vernacular name: –

Family: Percidae

Classification: intermediate

Distinguishing characteristics

The ruffe is a typical percid fish with ctenoid scales, an undivided dorsal fin with hard rays (spines) in the apical, and soft rays in the caudal section. Both pectoral and ventral fins are located on the anterior third of the body. Spines in D_1 XIV (XII–XV), D_2 12 (10 to 14), scales in the lateral line, 35 (34 to 37), gill rakers 8 to 12. The most typical characteristics distinguishing this species from *Gymnocephalus baloni* are the shape of the dorsal margin of the dorsal fin, which is oblique, and the shape of the operculum and sub-operculum, which have no serrations on the lower edge. For detailed information see Holčík and Hensel (1974). The ruffe is rather small, only occasionally reaching 20 to 25 cm. Lengths during the first to sixth year were recorded as 38, 62, 78, 90, 100, 112 mm, respectively. No striking differences in the growth rate have been recorded in Europe. For further information see Balon (1966), Oliva and Vostradovský (1960).

Distribution

The ruffe is found from England and northeastern France, as well as the Saone and Rhône Rivers, over Central to East Europe. In northern Europe it occurs in rivers flowing to the White Sea, in water courses eastwards as far as the Kolyma River and in the Caspian and Aral Sea Basins. It is not found in Ireland, Scotland, western and northern Norway, the Pyrenees and the Balkan Peninsula, in Italy or the Crimea. For details, see map, p. 286.

Ecology

This species occurs in the lower reaches of rivers and in man-made lakes. It does not require underwater plants, but prefers a clean bottom and avoids stretches where there are currents. It is considered a bottom dweller and bottom feeder, its diet consists of invertebrates, fish eggs and larvae according to season. It lives in schools.

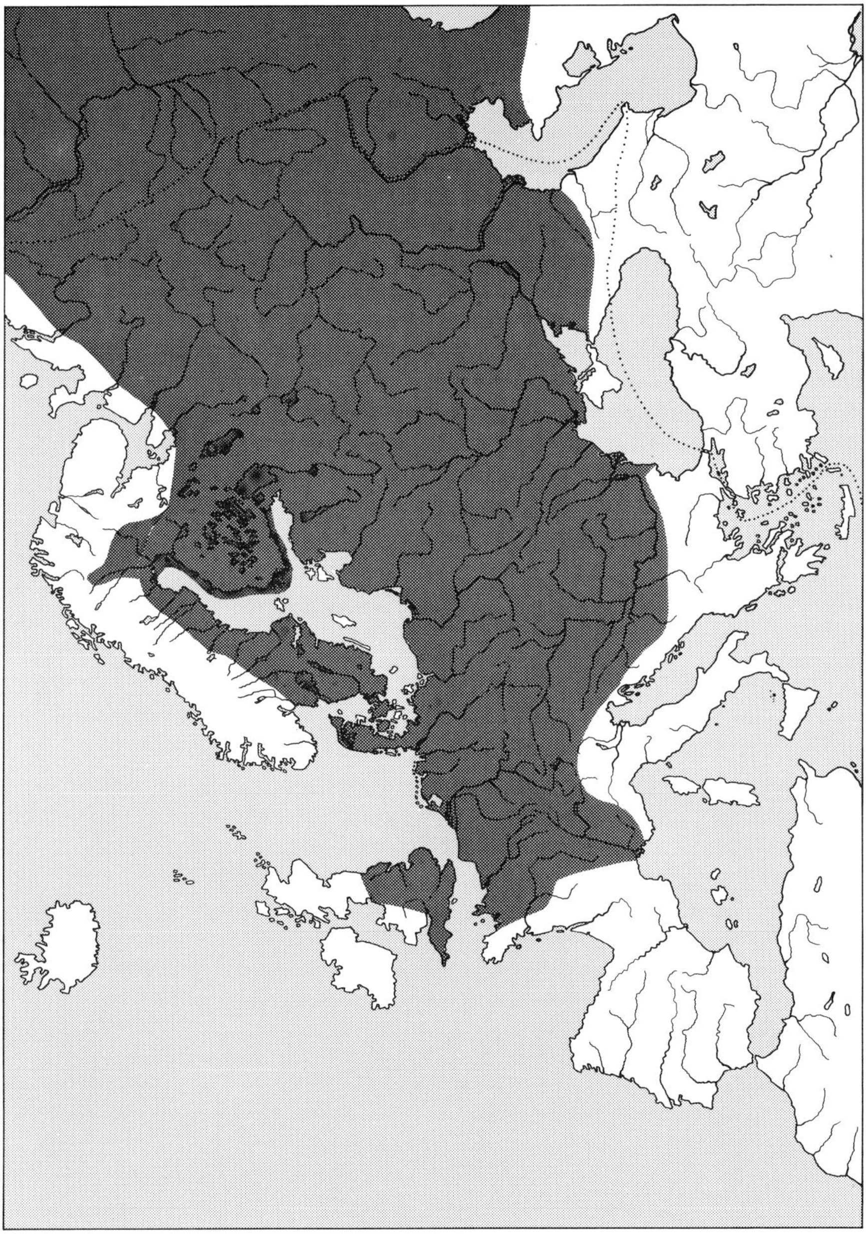

Fig. 75: European distribution of *Gymnocephalus cernuus* (Linnaeus, 1758).

Reasons for decline

It is difficult to speak of decline, as this species reflects the perturbed situation of many Central European waterbodies. The ruffe is often dominant in unbalanced populations of fishes, particularly in some man-made lakes, gravel-pits or larger, very slow flowing river reaches. When stratification of the waterbodies occurs, the ruffe dies out rapidly from a lack of oxygen in the lower water layers. However, its stock can recover quickly if the water quality improves.

Conservation measures

See general conservation measures of habitats in the lower reaches of streams, section 4, p. 316.

Gymnocephalus (Acerina) schraetser (Linnaeus, 1758)

E. Schraetzer
F. –
G. Schrätzer

Vernacular name: –

Family: Percidae

Classification: endangered

Distinguishing characteristics

The "Schrätzer" is a small fish, with all the typical patterns of the Percidae. It is easily distinguished from all other species of the genus *Gymnocephalus* by three to four dark, almost black, longitudinal stripes on the yellow sides. The dorsal fin bears dark oblique spots also arranged in regular longitudinal rows. D_1 XVII to XIX, D_2 12 to 13; 55–63 (65) scales in lateral line, subopercular bone serrated.

Distribution

It occurs in the Danube catchment area and in the little Kamchia River in Bulgaria.

Population in Europe

This species was never abundant, and it only occurs in the Danube and its large tributaries downstream from Bavaria. As this species has no commercial or sport value, no detailed information on its population is available.

Ecology

It is a rheophilic species usually found on the bottom in deep flowing water, where the substrate is gravel or sand and the current is swift. All available benthic invertebrates are taken as food. It never reaches a large size, the maximum recorded length being about 20 cm. Its activity increases during twilight and at night, when it reaches the shallows. It lives in small schools. Water quality requirements are high, and cool, well-oxygenated water seems to be a necessity for its survival.

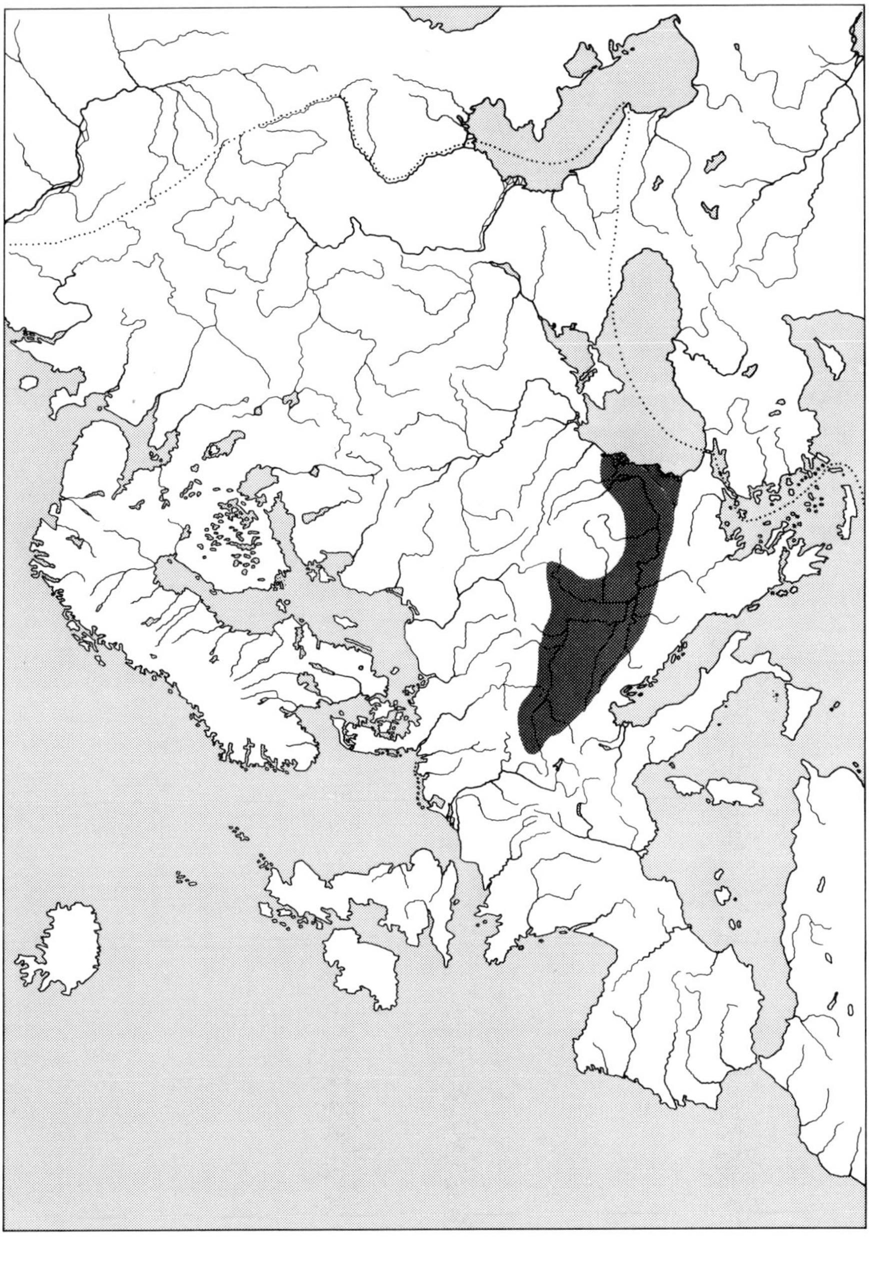

Fig. 76: Distribution of *Gymnocephalus schraetser* (Linnaeus, 1758).

Reasons for decline

Gymnocephalus schraetser seems to be sensitive to both habitat and water quality changes, which explains why its stock has diminished in many Danube tributaries in Germany. It disappeared from the Morava (March) River, one left-hand Danube tributary in Czechoslovakia, just after the Second World War, because of poor water quality and lack of sandy spawning grounds.

Conservation measures taken

No conservation measures have been taken so far.

Conservation measures proposed

As this species probably does not migrate, protection of habitats with large populations and water pollution control may be effective.

Romanichthys valsanicola Dumitrescu, Bănărescu and Stoica, 1957

E. –

F. –

G. Groppenbarsch

Vernacular name: Asprete (Romanian)

Family: Percidae

Classification: in the gravest danger

Distinguishing characteristics

This percid species bears, at first glance, a superficial resemblence to the bullhead (*Cottus* sp.) or a gudgeon. It has a low body, thicker anteriorly a large head wider than high. The eyes are large, the mouth, small. The maxillary is concealed posteriorly. The lateral line, scales ctenoid. D_1 IX, D_2 I, 15 to 16; lateral line 58 to 67 scales.

Distribution

This species is confined to the Argesul, Vislan and Riul Doamnei, three tributaries of the upper Arges, a tributary of the Danube in Romania. At present, it survives in the Vislan, upstream from its former range (Bănărescu, pers. comm.).

Population in Europe

This is a monotypic species confined to a very small area. When discovered and described in 1957 (Dumitrescu *et al.*) nobody expected it to approach extinction in a few years time (Stanescu 1971). Very probably it is now confined to one stretch of the Vislan where there are no villages (Bănărescu, pers. comm.).

Ecology

This species occupies habitats similar to those of the bullhead and stone loach, within the trout zone of this particular area of Romania. In mountain brooks, it hides under stones. Its food is mainly insects of the order Plecoptera, but other insects are also eaten.

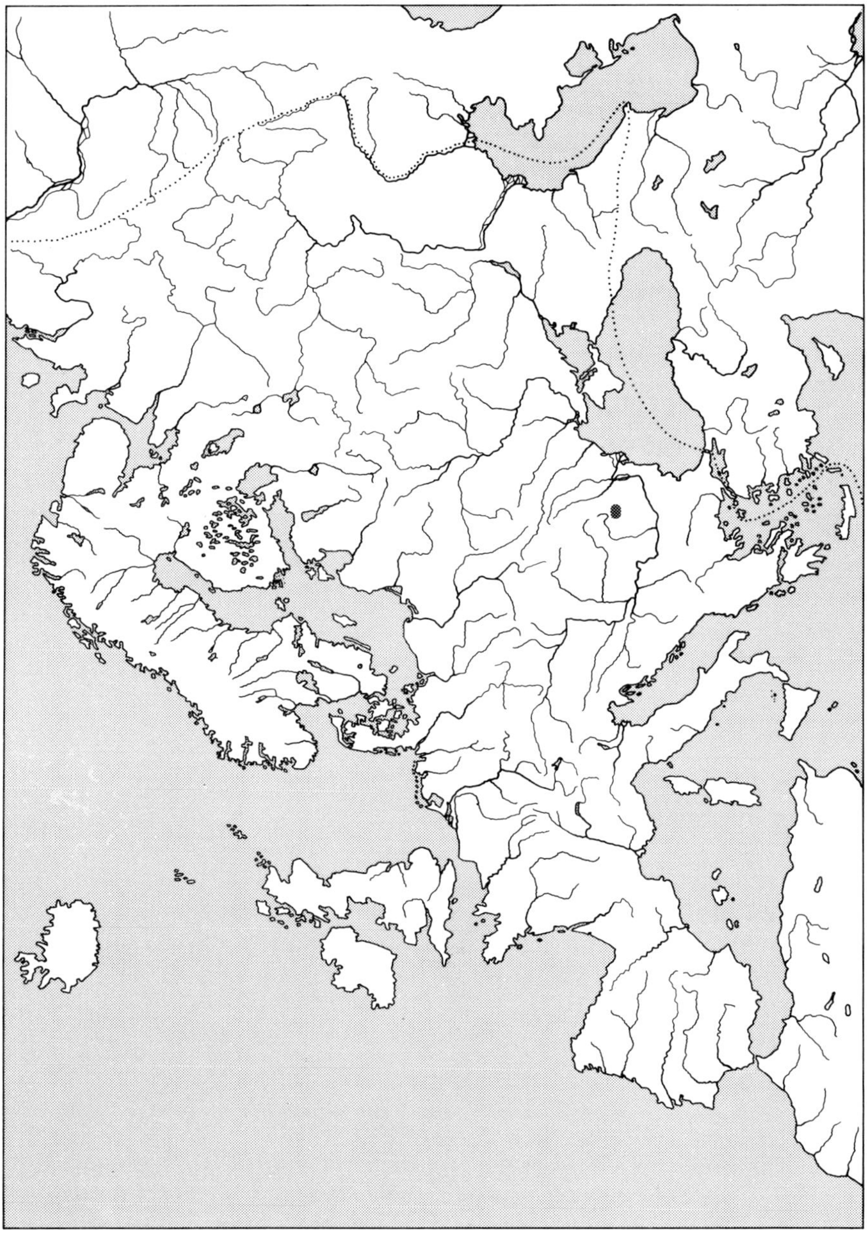

Fig. 77: Present distribution of the endemic species *Romanichthys valsanicola* (Dumitrescu, Bănărescu and Stoica, 1957) in Romania.

Reasons for decline

Decline is hardly the word to use; this species is almost extinct now. STANESCU (1971) stated that this is due to the direct influence of man-made habitat changes: construction of a dam, deforestation along the stream, road construction, repeated poisoning with herbicides and increased agrotechnical activity. Indirectly, it is influenced by the warmer water in stretches of deforested brooks and the spread of cyprinid species. STANESCU pointed out: "The combined action of all these factors, whose effects have become stronger and stronger since 1960, when construction of the dam began and the exploitation of forests was intensified, gradually, then rapidly, affected the biological balance determining the extinction of the most remarkable endemism of the Arges River drainage".

Conservation measures taken

This species became almost extinct before any conservation measures were taken. Recently, strict government regulations were introduced to protect the whole river stretch and its surrounding area.

Stizostedion lucioperca (Linnaeus, 1758)

E. Pikeperch
F. Sandre
G. Zander

Vernacular name: –

Family: Percidae

Classification: intermediate to vulnerable

Distinguishing characteristics

A typical representative of the Percidae with a rather elongated body. It often reaches a large size. Meristic data: D_1 XIII to XV (XVII); D_2 I to II (III), 19 to 24; 80 to 99 scales in the lateral line. The mouth contains many small teeth and two protruding canine teeth. The upper part of the body is dark olive green. The sides are lighter turning to silver on the belly. There are usually 8 dark and light vertical stripes on the sides. There is no pronounced sexual dimorphism, but during the spawning season, the males have dark spots on the belly.

Worldwide distribution

Like the perch, *Perca fluviatilis,* species of the genus *Stizostedion* are also found in the northern areas of North America. Worldwide five species are recognized:

Stizostedion canadense (Smith, 1834) known as the sauger (Fr. doré noir) and *Stizostedion vitreum* (Mitchill, 1818) known as the walleye (Fr. doré jaune) are confined to the American Continent. The three species *Stizostedion lucioperca, Stizostedion volgensis* and *Stizostedion marinum* have their original range in Europe. *S. marinum* is found only in small regions of the Black and Caspian Seas. This species is not discussed in this study, because no reliable information is available.

Distribution in Europe

The original range stretches from the Elbe River through the basins of the Baltic, Black, Azov, Caspian and Aral Seas to the Maritsa River which empties into the Aegean Sea. It is present throughout the Neva Basin, including Lakes Ladoga, Ilman and Onega. In the North (Finland), it reaches to 64° N and, sporadically, the Arctic Circle. This species has been introduced into the Rhine (1880) and its tributaries, and also into many

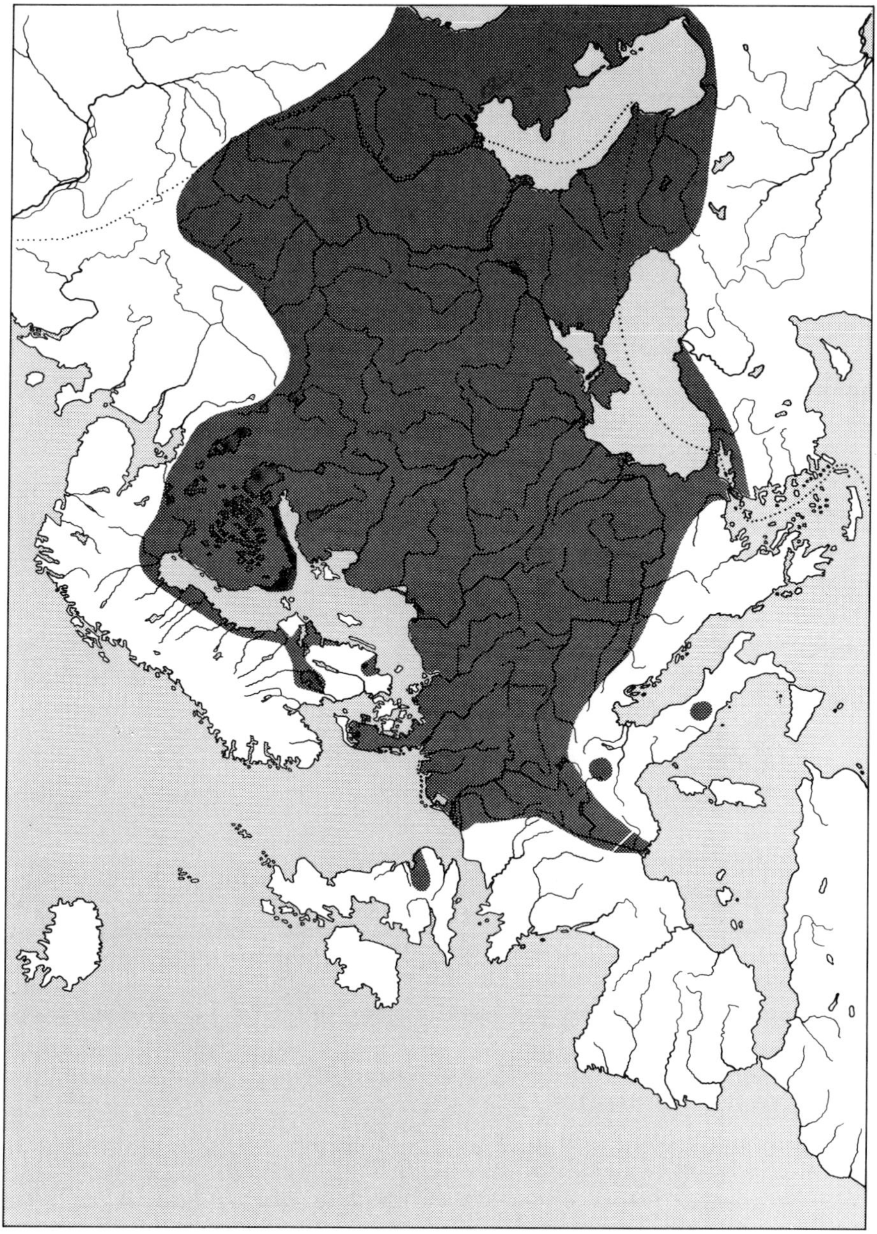

Fig. 78: Approximate european distribution of *Stizostedion lucioperca* (Linnaeus, 1758).

French rivers (GOUBIER, 1975), in the east-central part of England (MAITLAND, 1972), in Spain, Italy and Greece. On the distribution map, p. 295, the original range is shaded darker than the areas in the West, to which it was introduced.

Ecology

The pikeperch occurs in large rivers, adjacent lowland man-made lakes, and natural lakes. It can tolerate slightly turbid water, but it is rather sensitive to a low oxygen content. In large streams, it moves during the spawning time to adjacent backwaters or stays in relatively lentic sections of main rivers. Before the eggs are laid, the male prepares the nest about 0,5 m in diameter in rather shallow water, 0.5 to 1.5 m deep. The spawning ground is cleared of silt and deepened until the roots of submerged water plants protrude. The male is later joined by a female, which lays the eggs. The male guards the nest against attacking fishes and other animals and will even try to bite an approaching human hand! It washes silt out of the nest until the eggs are hatched. The fry mature at three to five years old, when it reaches a length of about 35 to 45 cm. It is an early spawner (April to the beginning of May) and during their first months, the young fishes begin to prey on the larvae of late spawners. Older fishes are typical predators and very useful for regulating populations of small cyprinid fishes in the lower reaches of rivers. It is highly prized commercially and is a sport fish, as well. Because of its spawning habits, it can be spread by transferring the eggs laid in artificial nests to other water bodies.

Reasons for decline

There has been no actual decline of pikeperch populations, because stocking is intensive enough to substitute for natural recruitment and mortalities in most of the "managed" waters. However, its situation could become critical if there were no artificial stocking.

Conservation measures taken

No active conservation measures have been taken to protect this species. Legal catch sizes and closed seasons, introduced in many countries throughout Europe, are not very effective. If an undersized individual is hooked, it will die soon after, even if treated very gently. In some countries, this has led to abolition of the size limit for anglers. Closed fishing seasons may prove very useful in natural lakes with commercial fisheries.

Conservation measures proposed

Protection of existing and creation of new both natural and artificial spawning sites would be the most effective measure.

Stizostedion volgensis (Gmelin, 1788)

E. Volga zander (pikeperch) — Vernacular name: –
F. –
G. Wolgazander — Family: Percidae

Classification: intermediate to vulnerable

Distinguishing characteristics

At first glance, this Volga pikeperch looks much like the common pikeperch, but it has however five to seven conspicuous dark stripes running vertically from the upper side over the sides to below the lateral line. It has no canine teeth, and is generally smaller than the common pikeperch, reaching a maximum of 40 cm in length and 1 kg in weight. Meristic features: D_1 XII to XIV, D_2 I to II, (19) 20 to 23; 70 to 79 (82) scales in the lateral line.

Worldwide distribution

See note on *Stizostedion lucioperca.*

Distribution in Europe

Rivers flowing to the Caspian and Aral Seas from the North, to the Black Sea and in the Danube. Its range reaches in the West to the confluence of the Morava (March) and Danube Rivers, near to Vienna. There are records

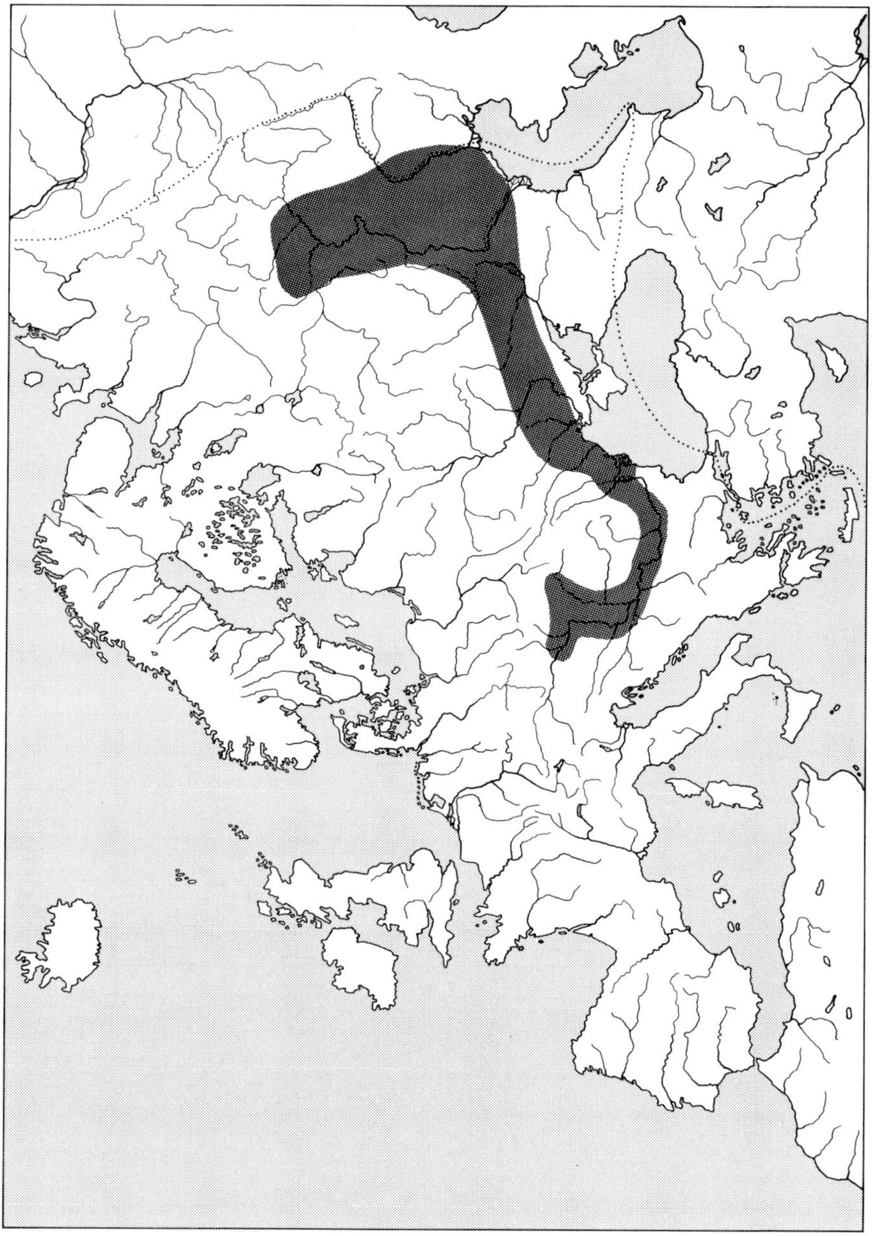

Fig. 79: Distribution of *Stizostedion volgensis* (Gmelin, 1788).

of its having been observed in the lowest reaches of the Váh River in Slovakia. Very little is known about its actual distribution.

Population in Europe

The lack of dependable information from the past and present suggests that this species has never been very abundant, at least not on the western part of its range. Even in the Volga Basin, it is not abundant, and its economic importance is minor.

Ecology

The ecology of this species is not fully known. BALON (1967) presumes it prefers the main river channel more than the common pikeperch. It spawns for the first time in its third or fourth year of life in April and May in biotopes with thick underwater vegetation. Its diet is similar to that of the common pikeperch.

Reasons for decline

In the main part of its range, it is less abundant now because the river banks have been altered and the suitable habitats are destroyed.

Conservation measures taken

Not known.

Conservation measures proposed

Fishery biologists should try to learn more about the ecology of this species so that effective protective measures may be taken.

Zingel zingel (Linnaeus, 1758)

E. Zingel
F. –
G. Zingel

Vernacular name: –

Family: Percidae

Classification: endangered

Distinguishing characteristics

Like most percid fishes the zingel has a strong, spiny first dorsal fin and a soft rayed second one, which is separated from the first. The pelvic fins are situated almost directly under the pectorals. The scales are ctenoid and present also on the head. The lateral line is conspicuous and extends to the end of the body. D_1 XI to XIV, D_2 I to II, 16 to 18; 82 to 95 scales in the lateral line. It has a greenish to brownish body with 4 irregular light and dark transverse stripes.

Distribution

This species occurs only in the Danube, the Dniester, and their tributaries; see distribution map, p. 301.

Population in Europe

The Zingel has never been found in abundance and is a rather rare fish. After the Second World War its abundance in the Danube and its tributaries in Germany, Austria and Slovakia decreased. It is very probable that its population has diminished further downstream, as well.

Ecology

The typical habitats are moderately deep or shallow, fast flowing streams with sand or gravel substrates. It feeds mainly on benthic invertebrates. It is most active during the night, when it comes closer to the shore. It is the largest species of this genus, reaching a maximum of 40 cm in length (average, 20 to 25 cm) and 250 g in weight. It is believed to hide its eggs as do the salmonids. In general, its ecology is not sufficiently known.

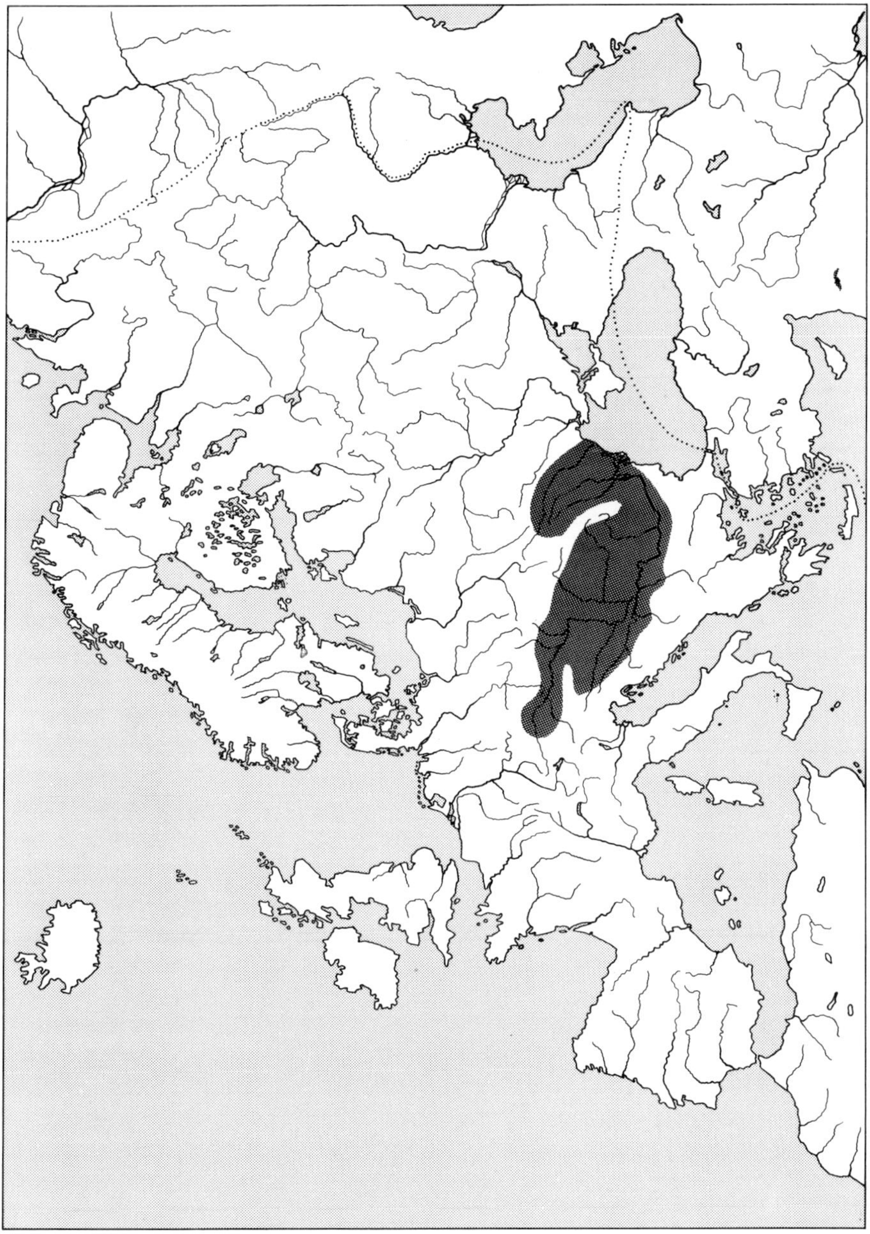

Fig. 80: Distribution of *Zingel zingel* (Linnaeus, 1758).

Reasons for decline

This species requires running, oxygen-saturated water. Man-made changes along the rivers, particularly damming and diminishing the velocity of the water flow, has resulted in deposition of silt in some river stretches. Changes in its habitat and increasing pollution have caused its scarcity.

Conservation measures taken

No effective conservation measures are known.

Conservation measures proposed

Protection of suitable lotic river stretches with the preferred substrate may help. Angling in these stretches should be banned. If sport fishing is allowed, then bait which lies on the bottom, such as earthworms, must be strictly forbidden.

Zingel asper (Linnaeus, 1758)

E. –
F. Apron
G. Rhône-Streber

Vernacular name: –

Family: Percidae

Classification: endangered

Distinguishing characteristics

The body of *Zingel asper* looks very similar to that of the closely related *Zingel streber.* The caudal peduncle is shorter, the colour, light brown to grey, with irregular transverse stripes. This species reaches only about 20 cm. D_1 VIII to X; D_2 I to II, 10 to 12; lateral line scales, 59 to 70, for further information, see OLIVA (1960).

Distribution

Z. asper is found only in the Rhône Basin. See map, p. 304.

Ecology

Presumably its habits and ecological requirements are similar to *Zingel streber.*

Conservation measures

See related species, particularly *Zingel streber.*

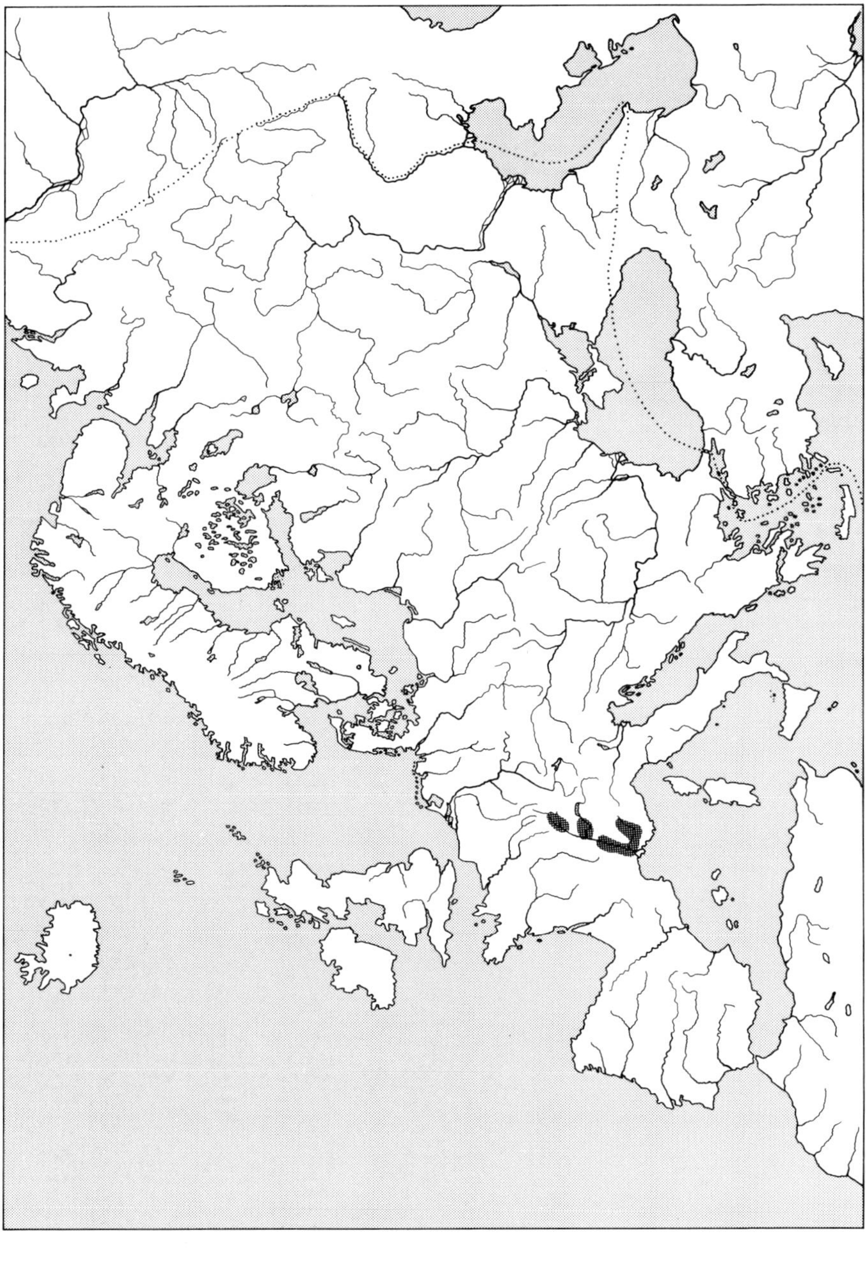

Fig. 81: Distribution of *Zingel asper* (Linnaeus, 1758).

Zingel streber Siebold, 1863

E. – Vernacular name: –
F. –
G. Streber Family: Percidae

Classification: endangered

Distinguishing characteristics

The general distinguishing characteristics of *Zingel zingel* also apply to *Zingel streber,* but the body of *Z. streber* is more slender and elongated, the caudal peduncle pencil-like. The colour of the body is a conspicuous yellow, with five dark transverse bands. The maximum length observed was about 20 cm. Meristic data: D_1 VIII to IX; D_2 I to II, 11 to 13; lateral line with 70 to 80 scales. From the Vardar, the subspecies *Zingel streber balcanicus* (Karaman, 1936) has been described. Its distinguishing characteristics are not conspicuous; for further information see KARAMAN (1936). The distribution of this subspecies is marked with a dotted line on the map for *Zingel streber.* A new subspecies *Zingel streber nerensis* Bănărescu Nalbant, 1979 is known from the Nera River, Romania.

Distribution

This species is typical of the Danube catchment area, including its tributaries. It has more or less the same range as *Zingel zingel,* but, being smaller, it is probably better adapted to fast flowing rivers and ascends further upstream. Its distribution area is thus larger than that of *Zingel zingel.*

Population in Europe

It has always been a rare species. It lives in small colonies or schools of several individuals, scattered within a suitable habitat. In general, its abundance is decreasing, particularly in the left-hand Danube tributaries.

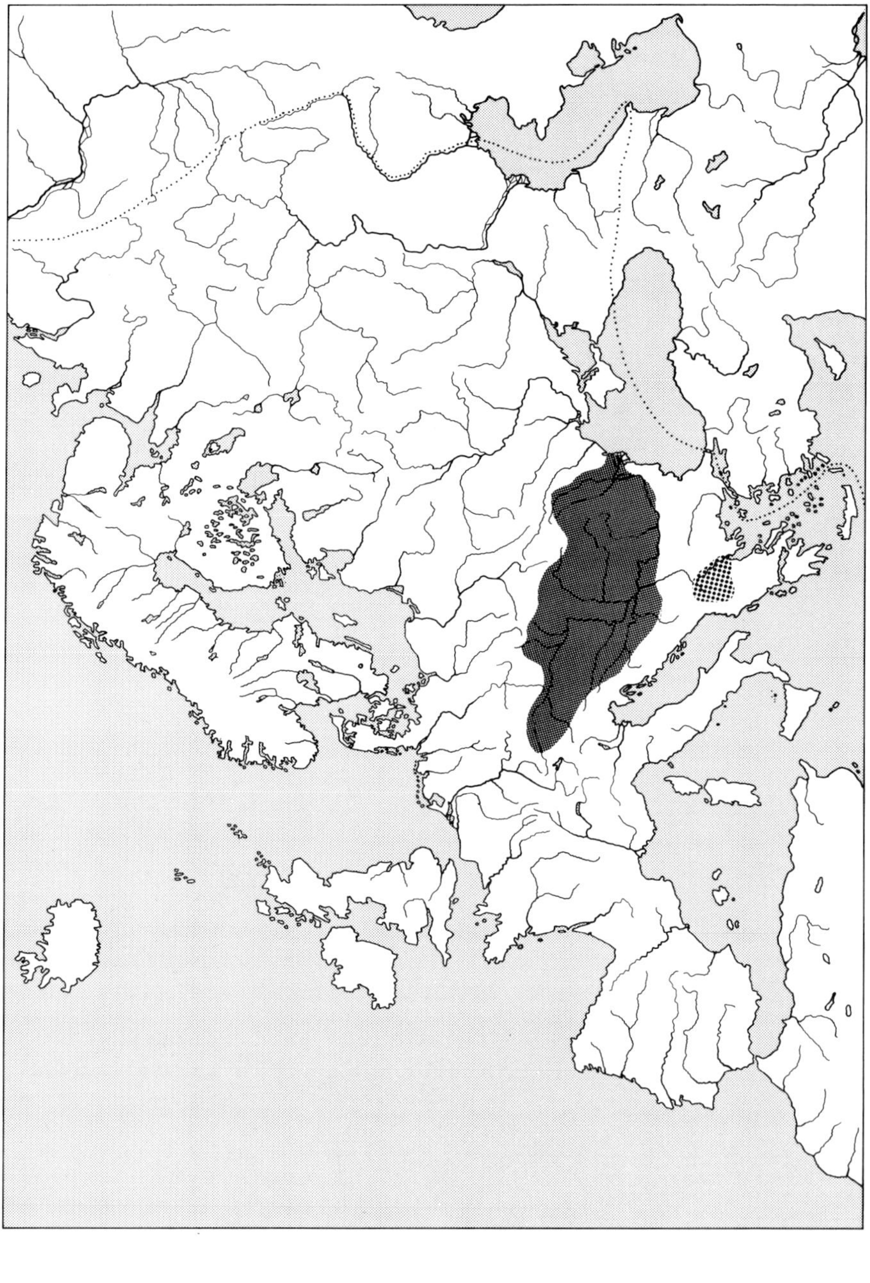

Fig. 82: Distribution of *Zingel streber* (Siebold, 1863), with subspecies *Zingel s. balcanicus* (see text, p. 305).

Ecology

Its ecology is comparable to that of *Zingel zingel,* but this species demands fast-flowing river stretches with sand or gravel substrate. Activity increases during the night, when it moves towards the shallow water.

Reasons for decline

Pollution and modification of the river bed, together with decreased and often fluctuating water discharge, are responsible for its decline. Changes in fish populations, particularly the increasing abundance of cyprinid fishes has very probably reduced its competition for survival.

Conservation measures taken

In some countries it is forbidden to catch this species. See *Zingel zingel.* No other conservation measures have been taken.

Conservation measures proposed

See *Zingel zingel.*

3.19. Gobiidae

Notes on the family Gobiidae

This family consists of seven genera and many species, of which only 15 are included in our list. They are primarily either salt water or brackish water species, but adaptable enough to live occasionally or permanently in freshwater. Those species living in freshwater are mainly found in the extreme eastern part of Europe and in the Mediterranean region. They are listed in the section 2.3. The family Gobiidae has little commercial value, particularly the species living in freshwater. This is the reason why there is no information on their previous occurrence or abundance that could be compared with recent records. It is strongly recommended that more attention be paid to these species, especially their abundance and life-cycles, as they seem to be sensitive to environmental alterations. In our study only two species are dealt with, the distribution of which extends far inland into Central Europe. One of them seems to be vulnerable to man-induced stress.

Gobius (Neogobius) fluviatilis Pallas, 1811

E. –

F. –

G. Fluß-Grundel

Vernacular name: Bytschok, babka (Russian)

Family: Gobiidae

Classification: rare

Distinguishing characteristics

It is a small species, reaching a maximum length slightly over 12 cm, displaying all characteristics of gobies. From other closely related species it may be distinguished as follows: the width of the head is identical for its hight or a little greater; the interorbital space is usually smaller than the eye diameter; the distance between the top of the rostrum and the anterior margin of the gill cover is identical to that from the posterior margin of orbit to the base of the first dorsal ray. In the second dorsal fin, the length of the rays decreases gradually toward the rear. The forehead to the edge of the orbits is covered by scales. The ventral fin usually extends to the anus; the caudal fin has 13 rays. Meristic data: D_1 VI to VII; D_2 I, 15 to 17.

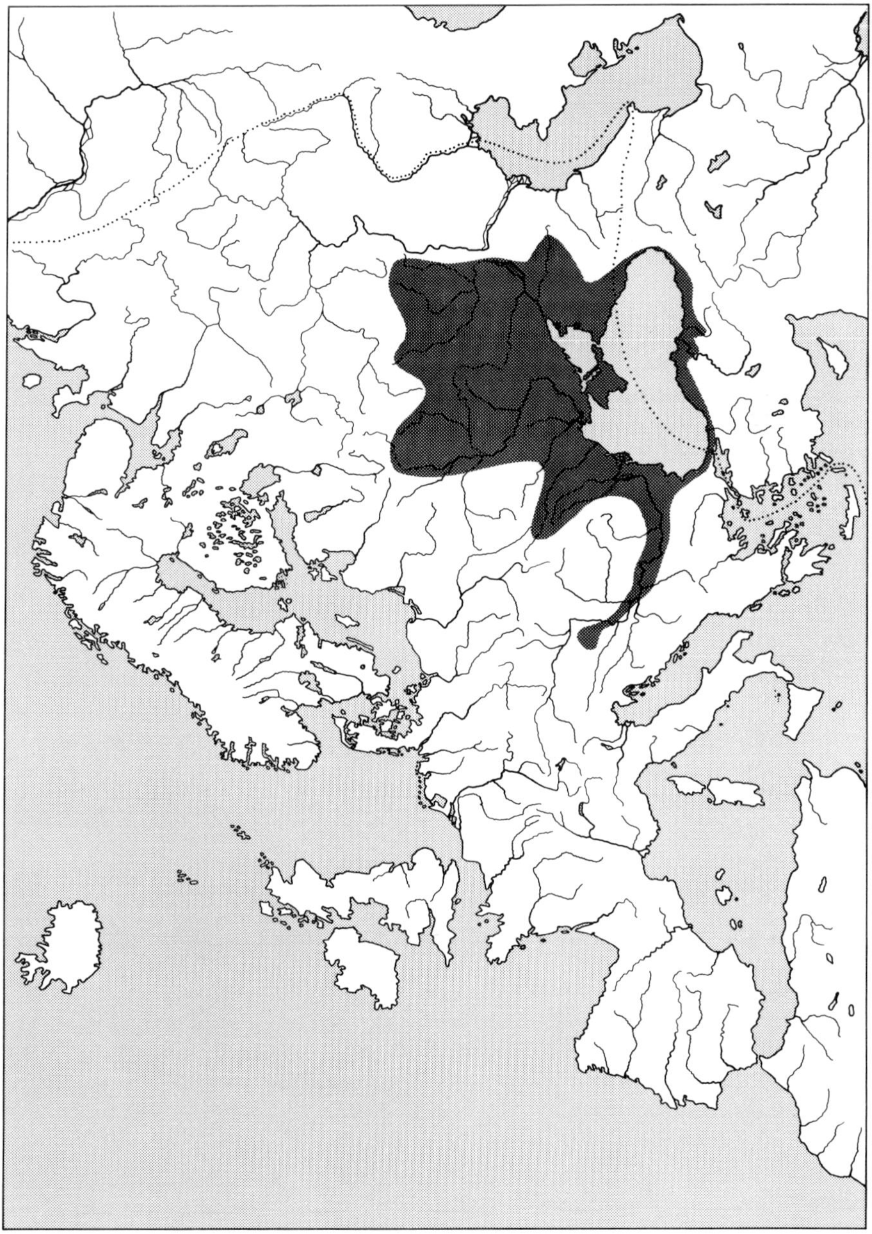

Fig. 83: Distribution of *Gobius fluviatilis* Pallas, 1811.

Distribution in Europe

It is known to occur from the lower Danube and the Black Sea Area and in the lower stretches of the Dnester, Bug, Dniepr, Desna, Don, Medvedica and Choper Rivers. The mouth of the Kuban, Crimean Rivers, the Porti and Batumi Regions, and along the coasts of Turkey, where it enters stretches of the Degirmendere, Kara Deresi, Sürmene and Ikizdere Rivers at high elevations. Its most inland occurrence is Lake Balaton, which is also the westernmost boundary of its range. See distribution map, p. 309 and Biro (1972).

Ecology

Gobius fluviatilis occurs mainly in coastal areas, particularly estuaries, but there are also purely freshwater populations. This goby has a short life span, of five to six years. Sexual maturity is reached in the second year. It is a benthic fish, living in both lentic and lotic waters. The oval eggs are deposited on stones. The male protects them for three to four days. It displays a pronounced sexual dimorphism. Males are dark during the spawning season. In lakes, this species may provide the food supply for a population of large predatory pikeperch.

Reasons for decline

There is some indication for recent decline of populations. This species may be sensitive to changes in water quality, especially chemical pollution. It seems to be particularly sensitive to alterations of the magnesium bicarbonate content. As this species was discovered in the eastern part of Central Europe only recently (since 1972), there are no data available for comparison to determine population fluctuations.

Conservation measures taken or proposed

No conservation measures are known. However, at first, improved knowledge on the distribution, the abundance and the ecological requirements of this species is necessary, for evaluation of the need of special conservation measures.

Proterorhinus marmoratus (Pallas, 1811)

E. – Vernacular name: –
F. –
G. Marmorierte Grundel Family: Gobiidae

Classification: vulnerable

Distinguishing characteristics

This small species, which reaches a maximum length of only about 10 cm, displays all the characteristics of gobies. It possesses no lateral line. The pelvic fins form one disc. There are two dorsal fins; the anterior one has spines, and the posterior, soft rays. The head is flat; the nostrils are extended into conspicuous tentacle-like tubes located above the lip of the upper jaw. Meristic data: D_1 VI to VII; D_2 I, 14 to 19.

Distribution in Europe

This species of gobies occurs in the Black and Caspian Sea Basins, both in salt and in brackish water. It also lives permanently in freshwater in the Danube, Bug and Dnieper Rivers. From the Azov Sea, it ascends the Don and Donets Rivers and it occurs in the Kuban River Delta. It is known from rivers flowing to the eastern Black Sea, as well as from the Maritsa River and its estuary and westernmost Turkey. It has been observed as far west as Neusiedler Lake in Austria and in Slovakian tributaries of the Danube including the Váh, Nitra, and Morava (March) Rivers.

Ecology

This originally rheophilic species has adapted itself to lentic environments, particularly canals and ponds densely overgrown with submerged vegetation. The first spawning takes place at two to three years. Most spent fishes die. The maximum life expectancy is probably five years. Spawning takes place in April and May. The spawning ground is a self-excavated depression or nest in shallow water; often the eggs are deposited in empty shells. It is a guarder of the eggs.

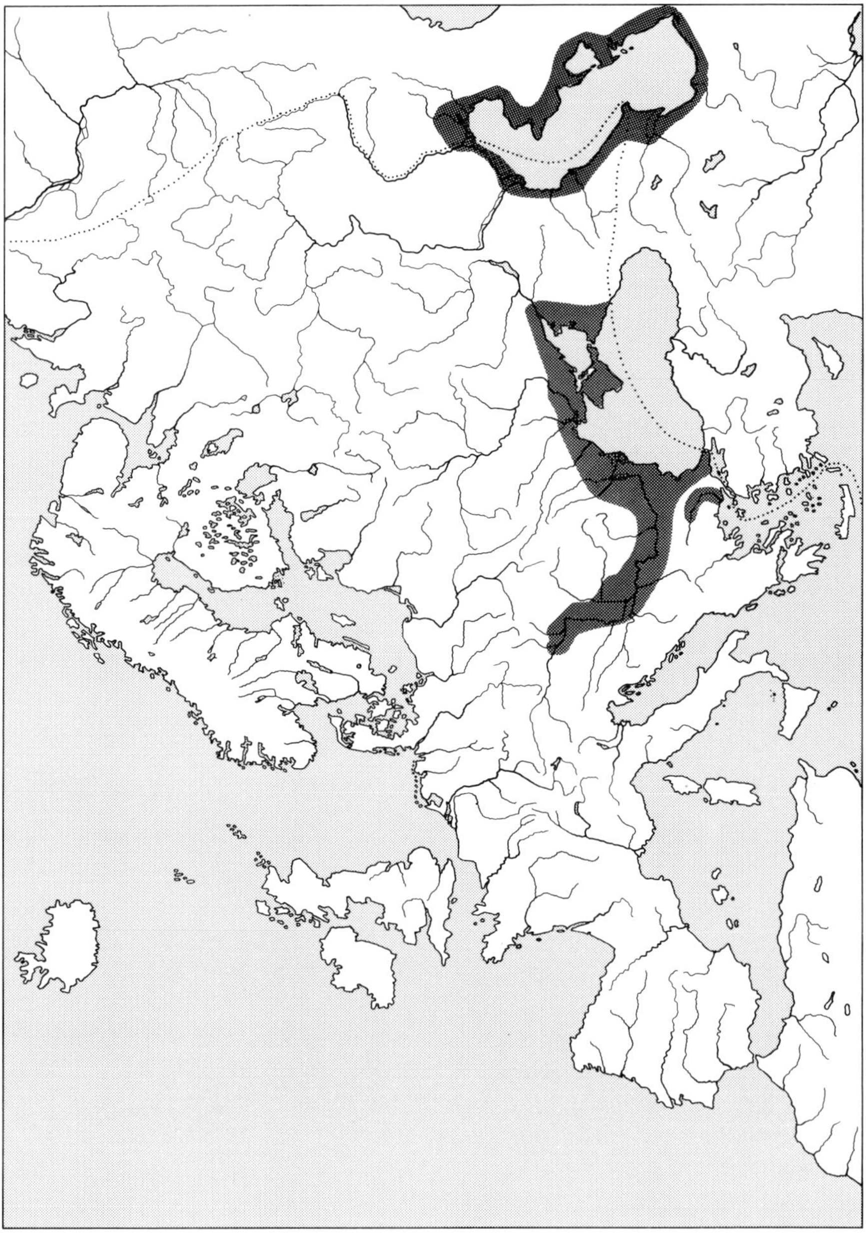

Fig. 84: Distribution of *Proterorhinus marmoratus* (Pallas, 1811).

Reasons for decline

In the westernmost part of its range, it was observed that habitat changes, particularly water level fluctuation, caused the disappearance of formerly abundant populations.

Conservation measures taken or proposed

No conservation measures are known. To conserve this species in its westernmost area of distribution, habitat protection is required. Alteration of the landscape, fluctuation in the water level, and changes in the vegetation must be avoided.

3.20. Pleuronectidae

Platichthys flesus (Linnaeus, 1758)

E. Flounder
F. Flet
G. Flunder

Vernacular name: –

Family: Pleuronectidae

Classification: rare to vulnerable in freshwaters

Distinguishing characteristics

This flat fish has eyes on the right side of the head, a prominent lower jaw and terminal mouth. Between the eyes, there is a sharp, bony keel reaching to the beginning of the lateral line. The body scales are small and smooth-edged, but on either side of the lateral line, and particularly between it and the pectoral fin, there are a few coarsely ridged scales, which give this region a rough feel. A single series of similar prickles along the bases of the dorsal and anal fins are the most noticeable identifying feature. The number of gillraker being 15 to 22 on the first gillarch distinguishes safely this species from the *Pleuronectes glacialis*. Meristic data: D 49 to 71; A 33 to 48, P 7 to 13.

Distribution in Europe

The population centers are in regions of the sea with low salinity in northern Europe: the Baltic and the North Sea. The general distribution includes European coasts from the White Sea to the Black Sea, as well as the Mediterranean. It is absent from Iceland. From the coast it penetrates far into freshwater.

Ecology

The flounder is a sea fish which prefers shallow water near estuaries along the European coast, especially in northern Europe. Young fishes penetrate into freshwater during the summer. Spawning takes place in salt water from January to April. The eggs are pelagic. The flounder takes benthic food.

Reasons for decline

Fluctuation in the stock is probably caused by heavy fishing and pollution of estuaries.

Conservation measures taken or proposed

Legal size limits and closed seasons have been introduced in many countries. It is hoped that water quality improvement in estuaries and better assessment of populations and fishery management will contribute to stabilizing the stocks.

Pleuronectes (Liopsetta) glacialis (Pallas, 1776)

E. – Vernacular name: Polarnaja kambala (Russian)

F. –

G. Polarscholle Family: Pleuronectidae

Classification: intermediate

Distinguishing characteristics

The body habitus is much like that of the flounder, but it is less elongated and bears no coarsely toothed scales, on the first gillarch 10 to 14 gillrakers. The males possess ctenoid, the females, cycloid scales. Meristic data: D 51 to 64, A 37 to 45.

Distribution and ecology

The 'Polarscholle' is found in the White and Barents Seas and in the coastal areas of northern Europe. It ascends such rivers as the Dvina. Molluscs are the main food.

Conservation measures taken or proposed

It is hoped that this species is not influenced by pollution. The only threat to its occurrence may be overfishing, which does not seem likely in this area.

4. Conservation proposals

The direct management of a fish species with a view to increasing its numbers, and consequently its production, in a given area or stretch of river is very often considered as the equivalent of protection and conservation. In this field, much work has been done to improve and maintain stocks of fish which may be harvested, but while this was being accomplished, the comprehensive protection of many indigenous species was either completely neglected or, more frequently, never considered at all. Instead of this merely practical approach, a new integrated effort is now required to protect and conserve the European fish fauna permanently, regardless of whether the species are small or large, palatable or unpalatable, provide good entertainment to the angler and fetch high prices on the market, or are unimportant economically. Conservation proposals presented here are very general, but may be useful as first approach to the problem. It is not possible to provide universal guide-lines that would apply to both northern and southern Europe, and to mountain, lowland and delta areas. It is intended to stress the necessity of environmental conservation, which is more important than the legal or administrative protection of a specific animal.

It must be stressed that comprehensive protection of the fauna is a prerequisite of a fishery practice. If this is neglected, it may well be that valuable genetic material will disappear forever. Living indigenous vertebrates, in this case fish, must be treated in the same way as any other cultural asset.

4.1. Utilitarian approach to conservation

The conservation and protection of species, as understood by both professional and sport fishermen, includes the care of the fishes: proper stocking of the water body, observance of the legal size limits and permitted fishing methods, and establishing of closed and open fishing seasons. In most European countries, these regulations are established and ruled by law.

No doubt, these measures have done a lot to maintain and preserve our indigenous fish species, such as the brown trout, grayling, pike, pikeperch, wels, carp, and tench, and above all they have generally encouraged the culture of fishes. Stocking is usually based on "trial and error" method. This approach has proved useful in waters inhabited by salmonids, for many years. In the last few decades, methods have been developed that provide an actual biological basis for determining the number of fish, that should be stocked (CUINAT et al. 1975). Stocking was recently rediscovered as a very useful method for maintenance of the fish population and for augmentation of rare and endangered species, both in Europe and in America (EIFAC / FAO 1982). This broadly supported and badly needed procedure is based on fishery biology, which seeks to produce workable models for both the temperate and tropical zones.

The best tool for the maintenance of fish stocks is the artificial stocking of natural water bodies. Fishery biologists have succeeded in producing fry of almost any species of fish required. As the construction of hatcheries is a rather expensive investment, large-scale enterprises able to cope with the high demands were preferred. This resulted in the stocking of waters with foreign populations. Brown trout for example were transported from one catchment area to another. In the case of coregonids, races endemic to one particular lake or small system of lakes were transferred to other water bodies. The hatchery procedures did not pay particular attention to the selection of parent fishes and release of fry into the waters of their origin. Cyprinid species, which are unable to reproduce themselves in artificial modified streams, were reared in ponds and released into rivers. Thus, to the casual observer the stock of fish often appears good. Moreover, the introduction of completely foreign species such as crucian carp, pumpkinseed, and catfish, has locally changed the fish fauna completely, while attempts to introduce other salmonids, Far Eastern cyprinids, Baikal grayling, the Asian *Hucho h. taimen,* and some pacific salmonids, are receiving more attention than the much-needed effort to let our indigenous fauna survive intact, at least in some selected places. The negative impact of *Micropterus salmoides* (large-mouthed bass, black-bass à grand bouche, persico trota) on the native small-sized fish fauna in southern Portugal should be considered as a warning against a procedure that should not be repeated.

To the credit of this utilitarian approach to fish protection, it must be said that many fishery biologists have proposed and carried out habitat improvements to remedy the lack of spawning substrate. Some of these scientists are ELSTER *et al.* (1973), VIBERT (1957), WHITE and BYNILDSON (1967), WHITE (1968). As the anglers and the few remaining commercial fishermen are constantly in contact with water and fish, they may be considered as the most vigilant guards against river pollution. And of course, dead fish floating upside down in a river make it clear even to the man in the street, that something is out of order in European rivers.

From the zoological viewpoint and from experience it is obvious, that man-made environments, artificially stocked with fish populations maintained for specific purposes, are only a poor substitute for a permanent protection and conservation of native fishes.

4.2. Conservation of species in whole sections of catchment areas

It is not always difficult to protect some selected species of fish. Where the fish requires only one kind of habitat for its whole lifespan (e.g. *Misgurnus fossilis* – Loche d'étang), conservation and protection is equivalent to habitat conservation. There are many possibilities for integrating the conservation of fish species with that of botanical "wetlands", bird sanctuaries or game reserves. The most important consideration is the maintenance of a stable ground water level. All landscape alterations must be avoided, and the use of herbicides and insecticides must be prohibited in the surroundings. Even if the former native species are missing from the chosen localities, artificial repopulation with indigenous species from other places may be employed.

The protection of the fish in the rhithral and the epirhithral zones does not seem to be too difficult. There are certainly enough small water courses in the underpopulated mountain areas of Europe available. One has only to choose an area with the required natural structure of the stream bed and its surroundings. No reservoirs for the collection of drinking water should be constructed anywhere nearby. Stretches of these streams are sometimes still in good hydrological and morphological condition. If not, environmental management may improve the habitat enough so that even endangered fish populations can recover. The species desired may also be reintroduced, if they have been eliminated. Even "small species", such as the bullhead and cyclostomes can be conserved in those areas.

The most complicated and urgent problem seems to be the protection and conservation of the species inhabiting the metarhithral, hyporhithral, and epipotamal zones. In Central Europe, these stretches, which are occupied primarily by large brown trout, grayling and barbel, are nearly all substantially changed or even destroyed in respect to their physical and biological characteristics. Practically all of them have been altered or subjected to various management practices. Most of these river stretches were altered in the past century to supply energy for saw and flower mills and to supply water for irrigation and carp ponds. At present, the water supply for small industries is no longer demanded, but along the water courses plenty of weirs, which fulfill mainly hydrological and regulating functions are still present and influence the ecosystems. It would be advisable to reconsider the expensive maintenance for the ancient weirs and to replace them, wherever possible, by more natural regulating barriers. Proposals to do so where made by LELEK (1982), and the positive benefits of modifications were documented by LUSK (1979), who provided also a list of references to this topic.

In addition to the leading species, many others occur there, and the large number of species require many types of ecological niches. LELEK and LUSK (1965) demonstrated, that fishes are more abundant where the bottom and shore lines are diversified. The ecosystem of fishes is often linked together into a food web. As the reproductive habits of the species differ, a variety of different spawning substrata must be provided. In many rivers, any attempt to restore the original physical characteristics of the bed, shore line, surrounding vegetation, and water quality would be absurd. Nevertheless, a river stretch satisfying the demands of the ontogenetic development of its species, including their local spawning migrations, must be found in each catchment area of our European hydrological systems. This section, which might be called a "river or fish sanctuary", must be in a condition still potential for regeneration, if the natural characteristics have already been altered. A river section with an inflowing tributary, representing a suitable area for implementation of successful conservation is shown in fig. 85. Furthermore, there must be a guarantee that no pollution will come from upstream and that no intensive agriculture will be practised there. Management, as it is usually understood (stocking, fishing, angling, removal of "coarse" fishes by electricity) must be abolished completely. Instead of "commercial" management, habitat improvement is required to provide the aquatic system with spawning grounds, hiding places for larvae, and nurseries for young fish, as well as feeding grounds for adult fishes. Deeper pools must also be available for the winter. The best guarantee for the maintenance of a variety of micro-habitats suitable along protected river stretches is the presence of a tributary. The difficulty in finding appropriate river sections would be greatest in the catchment areas of the Elbe, Rhine, Danube and Rhône Rivers, because those rivers are heavily influenced by man. Less difficulty is expected on the Iberian Peninsula, as the density of the human population is lower there. The selection of the appropriate river stretches, which need not be exceedingly long (estimated at 10 to 15 km of main stream and about 4 to 5 km of tributary), would be the task of national authorities. The best areas seem to be the frontier zones of Central European countries. Where zoogeographically important areas of the Balkan Peninsula are concerned, special arrangements would have to be made with the interested parties. Very encouraging indeed is the awareness of this issue in the Scandinavian countries and Great Britain (MAITLAND, 1979).

To protect and conserve the ecosystem of the potamal and particularly the hypopotamal (bream) zone, cannot be considered as wasted effort. Be-

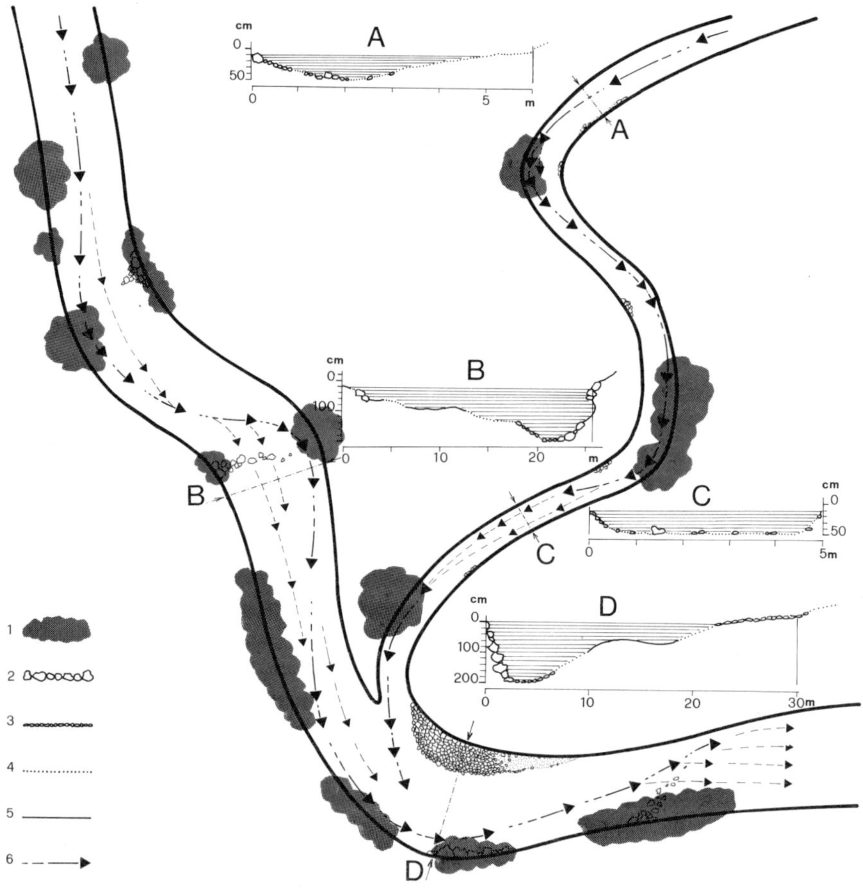

Fig. 85: An idealised stream section of the metarhithral (grayling to barbel zone), with an inflowing tributary.
1: shore vegetation, **2:** rock or stones, **3:** gravel, **4:** sand, **5:** sedimentation places, **6:** direction of water current.

cause these river stretches are mostly channelized to make them navigable and there is no hope of recovering the flood plains again, conservation effort must be focused on existing adjacent water bodies. In these stretches many phytophylic spawners are found, but not in the main stream because the water weeds are very scarce and the vegetation on the shoreline has either been removed or destroyed by the artificial wave action of passing ships. In these river sections, there are few backwaters, oxbows or dredged water bodies left. Provided these habitats have a free connection with the main stream, fishes can use them as spawning places. At present, occasionally flooded old backwaters provide the last spawning places for the "wild" form of the carp and other phytophylic species (LELEK 1978a, b, 1983). It would be advisable, before large unavoidable construction works are carried out, to include the creation of spawning areas in the planning. An indispensable source of Gasic information concerning the function in both natural and perturbated aquatic systems of running waters provides HYNES (1970).

Protection of the species occurring in lakes, particularly the coregonids and the charr, is difficult for several reasons. The fauna of the lakes where introgression has been observed can be saved only over the long term. Progressing euthrophication, linked with increased silting, may prevent or limit the natural spawning of coregonids. It seems that in the future, hatcheries will remain the inevitable supplementation and sometimes substitute for natural reproduction. If this is going to be the case, it would be most desirable to stock the lakes with fry produced by parent fishes from the same lakes or catchment areas.

Particular care must be given to the lakes on the Balkan Peninsula which are inhabited by a unique fish fauna.

4.3. Summary

To summarize this chapter on the actual protection that would help to rescue the rest of the indigenous fauna and make provisions for its future restoration, we can quote BUTCHER (1967), who suggested: ". . . the setting aside, difficult as it may be, . . . of environment reserves which must be complete in themselves, for example, a section of a watershed".

The last useful suggestion to all ichthyologists would be to make ichthyological nomenclature more accessible to all engaged environmentalists willing to contribute to the creation of "river reserves", and, last but not least, to speed up the effort to collect the needed ecological knowledge and life-cycle data on endangered or vulnerable taxa.

In this respect, this and the forthcoming volumes of "The Freshwater Fishes of Europe" shall help to a better understanding of the biology of the European fish fauna and as a first step to elaborate more effective methods of conservation and protection.

5. Acknowledgements

Many institutions, colleagues, and other persons have provided me with assistance in the form of advice, written communications, literature, and technical support. To all of them I express my sincere thanks.

Almaça, Prof. Dr., Carlos, NICC, Centro de Fauna Portuguesa das Universidades de Lisboa – PL2, Faculdade de Ciências – Museu Bocage, Rua Escola Politécnica, Lisboa 2, Portugal

Bănărescu, Prof. Dr., Petru, Ministerul Educatiei si Invatamintului, Institutul de Stinte Biologice, Splaiul Independentei, Bucuresti, Socialist Republic of Romania

Belgium: Ministère de l'Agriculture, Administration des Eaux et Forêts, Bruxelles

Brooks, R. J., Militant Association to save the Salmon, Vejle, Denmark and Laerdal, Norway

Canton de Vaud, Protection de la Nature et des Sites, Lausanne, Switzerland

Cervia, Prof. Dr. J. L., Museo Nacional de Ciencias Naturales C.S.I.C., Paseo de la Castellana 80, Madrid-6, Spain

Crossman, Prof. Dr. E. J., Royal Ontario Museum, Department of Ichthyology and Herpetology, Toronto, Canada.

Curry-Lindahl, Prof. Dr. K., UNEP-Nairobi and Department for Forestry and Resource Management College of Natural Resources, Univ. of California, Berkeley, California, U.S.A.

Dahl, Dr. Jørgen, Danmarks Fiskeri-og Havundersøgelser Ferskvandsfiskeri-laboratoriet, Silkeborg, Denmark

Denmark: Ministry of the Environment, National Agency for the Protection of Nature, Monuments and Sites, Copenhagen

Eidgenössisches Amt für Umweltschutz, Bern, Switzerland

Einarsson, Dr. E., Director of the Department of Botany, Museum of Natural History, Reykjavik, Iceland

Fahy, Dr. Edward, An Roinn Iascaigh, Department of Fisheries, Dublin, Ireland

France: Ministère de l'Environnement et du Cadre de Vie, Direction de la Protection de la Nature, Neuilly sur Seine

Frøiland, Dr. Øystein, Zoologisk Museum, Universitetet, Bergen, Norway

Finland: Ministry of Agriculture and Forestry, Bureau of Natural Resources, Helsinki

Federal Republic of Germany: Bundesministerium für Ernährung, Landwirtschaft und Forsten, Bonn

Greece: Ministère de Coordination, Service de la recherche scientifique et technique, Athenes

HARSANYI, Dr. A., Fachberater für Fischerei, Bezirk Niederbayern, Landshut, FRG
KÖLBING, Dr. A., Bayerische Landesanstalt für Fischerei, Starnberg, FRG
LEENTVAAR, Dr. P., Head of the Hydrobiology Department, Research Institute for Nature Management, Leersum, The Netherlands
LOTZ, Brigitte, Research Institute Senckenberg, Section Ichthyology II and Ecology of Fishes, Frankfurt, Federal Republic of Germany
Luxembourg: Administration des Eaux et Forêts, Service de l'Hydrobiologie
MAITLAND, Dr. P. S., Institute of Terrestrial Ecology, Wetlands Research Group, Edinburgh, Scotland
Malta: University of Malta, Msida
MICHAILOWA, Dr. Liljana, Bulgarian Academy of Science, National Natural History Museum, Sofia, Bulgaria
MÜLLER, Prof. Dr. K., University of Umea, Department of Biology, Section of Ecological Zoology, Sweden
National Agency for the Protection of Nature, Monuments and Sites; 13, Amaliegade, DK-1256 Copenhagen K, Denmark
SVÄRDSON, Prof. Dr. Gunnar, Director of the Institute of Freshwater Research, Drottningholm, Sweden
Sweden: The National Swedish Environment Protection Board, Solna
TORCHIO, Prof. Menico, Direttore Acquario Civico, via Gadio 2, Milano, Italy
Turkey: Ministry of Forestry, General Directorate of National Parks and Wildlife, Ankara
TUUNAINEN, Prof. Dr. Pekka, Finnish Game and Fisheries Research Institute, Helsinki, Finland
VUKOVIČ, Prof. Dr. T., Prirodno-matematicki fakultet, Sarajevo, Yugoslavia

6. Bibliography

ALM, G. A., 1959: Connection between maturity, size and age in fishes, Rep. Inst. Freshw. Res. Drottningholm 40: 5–145.

ALMAÇA, C., 1965: Contribution à la connaissance des poissons des eaux intérieures du Portugal. Separata da revista da Faculdade de Ciencias de Lisboa, 2 Ser. C. vol. **13** (2): 225–262.

ALMAÇA, C., 1972: Sur la systématique des Barbeaux (Genre et Sous-Genre Barbus) de la Péninsule Ibérique et de l'Afrique du Nord. Arg. Mus. Boc. (2.a série) Vol. **3**, No. 10: 319–346.

ALMAÇA, C., 1970: Sur les Barbeaux (Genre et Sous-Genre Barbus) de l'Afrique du Nord. Bull. du Mus. Nat. d'Hist. Naturelle, **2** Série 42 (1): 141–158.

ALMAÇA, C., 1976: Zoogeografia e especiacao dos Ciprinideos da Peninsula Ibérica. Soc. Portug. de Ciencias Nat., Col. Natura, vol. **4,** 28 pp., Lisboa.

ALMAÇA, C., 1983: Re-examination of the types of *Barbus haasi* Mertens, 1924 (Pisces: Cyprinidae). Senckenbergiana biol., **63**: 33–38.

ANDREASSON, S., 1971: Feeding habits of the sculpin (*Cottus gobio* L. Pisces) population. Rep. Inst. Freshw. Res., Drottningholm, **51:** 1–30.

APPLEGATE, V. C., 1950: Natural history of the sea lamprey, *Petromyzon marinus*. Michigan. U.S. Fish and Wildlife. Serv. Spec. Rep. **55.** 237 pp.

BACKIEL, T., and I. ZAWISZA, 1968: Synopsis of biological data on the bream *Abramis brama* (L.). FAO Fisheries Synopsis No. **36,** Rome.

BALON, E. K., 1964: Verzeichnis und ökologische Charakteristik der Fische der Donau. Hydrobiologica (Hague) **24**: 441–451.

BALON, E. K., 1967: Fishes of Slovakia. Obzor, Bratislava (In Slovak).

BALON, E. K., 1967: Evolution of the Danube fish fauna, its recent state and an attempt for the prognosis of further changes after the hydroelectric power buildings. Biological Works (Bratislava) **13:** 121 pp. (In Slovak with summaries in English, German, an Russian.)

BALON, E. K., 1968: Urgeschichte der Donau-Ichthyofauna (vor dem Einfluss seitens des Menschen). Arch. Hydrobiol. Suppl. **34:** 204–222.

BALON, E. K., 1969: Ein Versuch, durch die Analyse der Herkunft und der historischen Entwicklung von Forellen die Frage zu beantworten: Sind diese Fische aus dem Donau- und Weichselstromgebiet unterscheidbar? Zbornik Vych. Slov. Mus. Ser. B (**10**): 23–40 (Slovak with summary in German).

BALON, E. K. (Ed.) 1980: Chars–Salmonid Fishes of the Genus *Salvelinus*. Dr. Junk Publ., The Hague, 928pp.

BALON, E. K., S. S. CRAWFORD & A. LELEK, 1987: Is the sudden occurrence of *Gobio albipinnatus* Lukasch, 1933 in the upper Danube a result of upriver inva-

sion or sympatric speciation (Pisces: Cyprinidae)? Senckenbergiana biol (In press).
BALON, E. K., S. S. CRAWFORD & A. LELEK, 1987: Are there sympatric forms of *Vimba* Fitzinger 1873 in the Danube near the future connection to the Main River? (Pisces, Cyprinidae). Senck. biol. (In press).
BĂNĂRESCU, P., 1964: Fauna Republici populare Romine, Pisces-Osteichthyes. Vol. **13,** Acad. R.P.R., Bucuresti.
BĂNĂRESCU, P., 1965: Das Becken der unteren Donau als eiszeitliches Rückzugsgebiet der Fischfauna. Arch. Hydrobiol. Suppl. **30:** 24–35.
BĂNĂRESCU, P., 1975: Principles and Problems of Zoogeography (Principii si probleme de Zoogeografie). Translated from Roumanian, publ. for U.S. Dept. of Commerce & Nat. Sci. Foundation, Washington D.C., NOLIT Publ. House, Belgrade, 214 pp & 2 appendices.
BĂNĂRESCU, P., M. BLANC, J. L. GAUDET and J. C. HUREAU, 1971: European Inland Water Fish. A multilingual catalogue. Fishing News (Books) Ltd. London.
BAUCH, G., 1955: Die einheimischen Süßwasserfische. 3rd ed. Numan Verl. Radebeul and Berlin.
BERG, L. S., 1933: Übersicht der Verbreitung der Süsswasserfische Europas. Zoogeographica **1:** 107–208.
BERG, L. S., 1948–1949: Freshwater fishes of the USSR and adjacent countries. Vol. **1–3** (Translated by Israel Programme for Sci. Transl. Jerusalem).
BIANCO, P. G., 1980: Areale Italico rinvenimento in Calabria e origini delle popolazioni Mediterrance di *Gasterosteus aculeatus* L. Boll. Mus. Civ. St. Nat. Verona **7:** 197–216.
BIRO, P., 1972: *Neogobius fluviatilis* in Lake Balaton – a Ponto-Caspian goby new to the fauna of Central Europe. J. Fish. Biol. **4:** 249–255.
BLESS, R., 1982: Untersuchungen zur Substratpräferenz der Groppe, *Cottus gobio* Linnaeus 1758. Senckenbergiana biol. **63:** 161–165.
BORNE, M. v. d., 1877: Wie kann man unsere Gewässer nach den in ihnen vorkommenden Arten klassifizieren? Cirkular des Deutschen Fischereivereines **4.**
BRUNKEN, H., 1987: Ausbreitungsdynamik von *Noemacheilus barbatulus* (Linnaeus, 1758). Ph. D. Thesis, Dept. of Zoology, Univ. of Braunschweig, Fed. Rep. of Germany.
BUTCHER, A. D., 1967: A changing aquatic fauna in a changing environment. IUCN Publ., n. s. **9:** 197–218.
COLLARES-PEREIRA, M. J., 1978: *Chondrostoma toxosoma arrigonis* (Steindachner, 1866), a new cyprinid to Portugal. Arg. Mus. Boc. **2:** 219–239.
COLLARES-PEREIRA, M. J., 1983(a): Les Phoxinelles circum-méditerranéens (avec la description d'*Anaecypris* N. Gen.) (Poissons, Cyprinidae). Cybium, 7 (2): 85–91.
COLLARES-PEREIRA, M. J., 1983(b): Estudo sistemático e citogenetico dos pequenos ciprinídeos Ibeéricos pertencentes aos géneros *Chondrostoma* Agassiz, 1835, *Rutilus* Rafinesque, 1820 e *Anaecypris* Collares-Pereira, 1983. Dissertaçào apresentada à Faculdade de Ciències da Universidade de Lisboa, Lisboa, 1983: 511 pp.
COLLETTE, B. B., and P. BĂNĂRESCU, 1977: Systematics and zoogeography of the fishes of the family Percidae. J. Fish. Res. Board Can. **34:** 1450–1463 (Percis-Symposium).

Coubier, J., 1975: Biogéographie, biométrie et biologie du sandre *Lucioperca lucioperca* (L.) Osteichthyes, Percidae. Ph. D. Thesis. Univ. Claude Bernard, Lyon (ex Colette and Bănărescu, 1977).

Cuinat, R., J. A. Timmermans, J. Arrignon and G. Tufféry, 1975: Ecological diagnosis in salmonid streams; methods and example. EIFAC Techn. Pap. **22:** 133 pp.

Dumitrescu, M., P. Bănărescu and N. Stoica, 1957: *Romanichthys valsanicola,* Nov. gen. sp. (Pisces: Percidae). Trav. Mus. Hist. Nat. "Grigore Antipa" **1:** 225–244.

Duncker, G., 1960: Die Fische der Nordmark, revised by W. Ladiges, Hamburg.

Deelder, C. L., 1970: Synopsis of biological data on the eel *Anguilla anguilla* (Linnaeus, 1758). FAO Fisheries Synopsis No. **80.** FAO, Rome.

EIFAC/FAO, 1982: Report of the Symposium on stock enhancement in the management of freshwater fishes. EIFAC Techn. Pap. **42.** 43 pp.

Elster, H.-J., D. P. Huber, E. Rehbronn and O.-K. Trahms, 1973: Vorschläge zum Schutz der Fischerei beim Gewässerbau. Arch. FischWiss. **24.** 28 pp.

Ferguson, A., K. J. M. Himberg and G. Svardson, 1978: Systematics of the Irish pollan (*Coregonus pollan,* Thompson): an electrophoretic comparison with other holarctic Coregonidae. J. Fish. Biol. **12:** 221–233.

Frič (Fritsch), A., 1872: Die Wirbeltiere Böhmens. Arch. f. naturwiss. Landesdurchforschung von Böhmen, 2 (2). Prague. 152 pp.

Frič (Fritsch), A., 1888: Vysvetlivky ku rybarske mape Čech. (The legend to the fish distribution map of Bohemia), Prague.

Gasowska, M. (Ed.), 1962: Klucze do oznaczania kregowcow Polski, Part I, Pisces, Krakow.

Gaumert, D., 1981: Süßwasserfische in Niedersachsen, Arten und Verbreitung als Grundlage für den Fischartenschutz. Ministry of Food, Agriculture and Forestry, Hannover, Fed. Rep. of Germany.

Haempel, O., 1930: Fischereibiologie der Alpenseen, VIII. Schweizerbart'sche Verlagsbuchhandlung Stuttgart.

Harsányi, A., 1982: Der Huchen. Paul Parey. Hamburg.

Harsányi, A., 1986: Huchen – Hucho hucho (L. 1758) – eine bedrohte Fischart. Schützt die Gewässer – Petri-Stiftung, **2:** 13–16.

Hartmann, J. & W. Nümann, 1977: Percids of Lake Constance, a Lake Undergoring Eutrophication. J. Fish. Res. Board Can. **34:** 1670–1677.

Hensel, K., 1971: Some notes on the systematic status of *Carassius auratus gibelio* (Bloch, 1782) with further record of this fish from the Danube River in Czechoslovakia. Acta soc. zool. Bohemoslov. **35:** 186–198.

Hensel, K., 1979: *Rutilus (Pararutilus) frissi meidingeri* in Czechoslovak stretch of the Danube River. Acta soc. zool. Bohemoslov. 42: 250–252.

Hensel, K., and J. Holčík, 1983: On the identity of *Hucho hucho* and *Hucho taimen* (Pisces: Salmonidae). Folia Zoologica 32: 67–83.

Heuschmann, O., 1957: Die Weissfische (Cyprinidae). Im Handbuch der Binnenfischerei Mitteleuropas, Bd. **III. B:** 23–199, Schweizerbart'sche Verlagsbuchhandlung, Sutttgart.

Holčík, J., 1982: Review of experiments with introduction and acclimatization of the huchen – *Hucho hucho* (Linnaeus, 1758), (Salmonidae). EIFAC Symposium on stock enhancement in the management of freshwater fisheries. Budapest, May 31–June 5, 1982.

Holčík, J., and K. Hensel, 1972: Handbook of Ichthyologie (Ichthyologicka prirucka). Obzor, Bratislava.

Holčík, J., and K. Hensel, 1974: A new species of *Gymnocephalus* (Pisces: Percidae) from the Danube, with remarks on the Genus. Copeia: 2, 471–486.

Holčík, J., K. Hensel, J. Nieslanik & S. Skácel, 1984: Hlavátka, Hucho hucho (Linnaeus, 1758); (The huchen, Hucho hucho (Linnaeus, 1758), Veda, Bratislava. (In Slovak).

Huet, M., 1949: Aperçu des relations entre la pente et les populations piscicoles des eaux courantes. Rev. Suisse d'Hydrol. **11:** 411–432.

Huet, M., A. Lelek, J. Libosvárský and M. Peňáz, 1969: Contribution à l'identification des zones piscicoles de quelques cours d'eau de Moravie (Tchécoslovaquie). Verh. Internat. Verein. Limnol. **17:** 1103–1111.

Illies, J., 1961: Versuch einer allgemeinen biozönotischen Gliederung der Fließgewässer. Int. Rev. ges. Hydrobiol. **46:** 205–213.

Illies, J., 1961a: Gebirgsbäche in Europa und Südamerika – ein limnologischer Vergleich. Verh. int. Verein Limnol. **14:** 517–523.

Jungwirth, M., 1978: Some notes to the farming and conservation of the Danube salmon *(Hucho hucho)*. Env. Biol. Fish. **3:** 231–234.

Karaman, S., 1936: 10. Beitrag zur Kenntnis der Süßwasserfische Jugoslawiens. Bull. Soc. Scient. Skoplje. **17:** 55–63.

Klausewitz, W., 1974: Die frühere Fischfauna des Untermains. Natur und Museum **104:** 1–7.

Klemetzen, A., 1973: Pelagic, plankton-eating perch. Astarte (Oslo) **6:** 27–33.

Koli, L., 1977: Retkeilijan kalaopas. Helsinki.

Krefft, G., and H. Michaelis, 1976: Die Meeräsche im niedersächsischen Wattenmeer. Natur und Museum **106:** 23–29.

Krupka, I., and J. Holčik, 1976: On the occurrence of *Barbus plebejus* Bonaparte, 1832 in the Poprad River (Vistula Basin, Czechoslovakia) with regard to its assumed hybrid origin. Acta soc. zool. Bohemoslov. **40:** 163–178.

Kux, Z. and J. Libosvárský, 1957: Zur Verbreitung und Rassenzugehörigkeit der europäischen Hundsfische (*Umbra krameri* Walbaum 1792 = *lacustris* Gross. 1794). Zool. listy **6** (20) 3: 215–224.

Ladiges, W., 1960: Süßwasserfische der Türkei, 1. Teil Cyprinidae. Mitt. Hamburg Zool. Mus. Inst. **58:** 105–150

Ladiges, W., and D. Vogt, 1965, 1979 (2. A.): Die Süßwasserfische Europas bis zum Ural und Kaspischen Meer. Paul Parey, Hamburg and Berlin.

Lauterborn, R., 1916, 1918: Die geographische und biologische Gliederung des Rheinstroms, I and III. Teil. Sitzungsber. d. Heidelb. Akad. d. Wiss., Math.-naturwiss. Klasse, Abt. B. Heidelberg, C. Winters Universitätsbuchhandlung.

Lelek, A., 1973: Occurrence of the sea lamprey in midwater of Europe. Copeia **1973, 1:** 136–137.

Lelek, A., 1975: Erster fischereibiologischer Einsatz des Forschungsbootes "Courier" im Main und Rhein. Natur und Museum **105:** 312–316.

LELEK, A., 1977: Perish in silence. Naturopa **28:** 13–18.

LELEK, A., 1978a: Die Fischbesiedlung des nördlichen Oberrheins und des südlichen Mittelrheins. Natur und Museum **108:** 1–9.

LELEK, A., 1978b: Die Bedeutung der Altrheine für die Fischfauna des Rhein-Hauptstromes am Beispiel des Schusterwörther Altrheins. In: Ökologische Modelluntersuchungen "Schusterwörther Altrhein", Courier Forschungsinstitut Senckenberg **35:** 109–154.

LELEK, A., 1980a: Einige Notizen zum Schutz der Süßwasserfische in der Bundesrepublik Deutschland. Natur und Landschaft **55:** 295–298.

LELEK, A., 1980b: The occurrence, taxonomy and future of trouts in north-east Turkey. Council of Europe, Eur. Committee for the Conservation of Nature and Natural Resources. CDSN (81) **11.** 110 pp.

LELEK, A., 1981: Das Fischvorkommen in den unteren 20 km des Nidda-Flußlaufes. (Eine Bestandsaufnahme, sowie Vorschläge zur Regenerierung der Fischfauna dieser Flußstrecke und deren Nebengewässer. Stadtentwässerungsamt, Auftr. Nr. 080 680, Stadt Frankfurt am Main.

LELEK, A., 1983: Ichthyologische und fischereibiologische Arbeiten in den hessischen Altrheinen. Naturschutz und Landschaftspflege in Hessen 1982/83. Der Hessische Minister für Landesentwicklung, Umwelt, Landwirtschaft und Forsten (Wiesbaden).

LELEK, A., and S. LUSK, 1965: Occurrence of fishes in relation to formation of the bed. Zool. listy (Folia Zoologica) **14:** 225–234. (Czech with summary in English).

LELEK, A., and M. PEŇÁZ, 1963: Spawning of *Chondrostoma nasus* (L.) in the Brumovka River. Zool. listy (Folia Zoologia) **12:** 121–134.

LELEK, A., J. LIBOSVÁRSKÝ, M. PEŇÁZ, R. BEZDĚK and Z. MACHÁČEK, 1964: Observation on Fish under Ice in Winter. Ecologia Polska, A (12–16): 305–312.

LIBOSVÁRSKÝ, J., 1957: On the ecology and reproduction of the stone-loach *Nemacheilus barbatulus* (L.). Zool. listy (Folia Zoologica) **6** (20) 4: 367–388. (Czech with summaries in English and Russian).

LIBOSVÁRSKÝ, J., 1961: Zur palaeoborealen Verbreitung der Gattung *Carassius* (Jarocki, 1822). Zool. Jb. Syst. **90:** 197–210.

LILLELUND, K., 1961: Untersuchungen über die Biologie und Populationsdynamik des Stintes, *Osmerus eperlanus* (L.), der Elbe. Archiv FischWiss. **12:** 1–127.

LUSK, S., 1967: Population dynamics of *Chondrostoma nasus* (Linnaeus, 1758) in the Rokytna River. Acta sc. nat. Brno **12:** 475–522. Academia, Prague.

LUSK, S., 1979: Rocky chutes and the fish stock of streams. Acta Sc. Nat. Brno **13:** 1–26.

LUSK, S., V. BARUŠ, and V. VESELÝ, 1977: On the occurrence of *Carassius auratus* in the Morava River drainage area. Folia Zoologia **26:** 377–381.

MAITLAND, P. S., 1972: A key to the freshwater fishes of the British Isles with notes on their distribution and ecology. Freshw. Biol. Assoc. No. 27.

MACCRIMMON, H. R., 1971: World distribution of rainbow trout *(Salmo gaidneri)*. J. Fish. Res. Board Canada **28:** 663–704.

MOHR, E., 1941: Stint *(Osmerus artedi)*, Lieferung 7: 503–526. In "Handbuch der Binnenfischerei Mitteleuropas", Band **III A,** Schweizerbart'sche Verlagsbuchhandlung, Stuttgart.

MOHR, E., 1952: Der Stör. Die Neue Brehm-Bücherei, Leipzig.

MÜLLER, K., 1960: Beitrag zur Systematik und Verbreitung von *Cottus gobio* L. und *Cottus poecilopus* Heckel. K. Fysiogr. Sällsk. Lund Förhandl. **30:** 57–66.

NELSON, J. S., 1976: Fishes of the World. J. Wiley, New York, London, Sydney, Toronto.

NERESHEIMER, E., 1937: Die Lachsartigen (Salmonidae). I. Teil: 219–370. In "Handbuch der Binnenfischerei Mitteleuropas", Band **III A,** Schweizerbart'sche Verlagsbuchhandlung, Stuttgart.

NIKOLSKI, G. W., 1957: Spezielle Fischkunde. Berlin. (German translation).

NOWICKI, M., 1882: Krainy rybne Wisly (The area of fish distribution in the Wisla River). Gazeta rolnicza **23:** 24.

NÜMANN, W., 1973: Versuch einer Begründung für den Wandel in der qualitativen und quantitativen Zusammensetzung des Fischbestandes im Bodensee während der letzten 60 Jahre und eine Bewertung der Besatzmaßnahmen. Schweiz. Z. Hydrobiol. **35:** 206–238.

OLIVA, O., 1959: A note on *Acerina acerina* (Güldenstädt). Acta. Univ. Carolinae – Biologica **3:** 165–168.

OLIVA, O., 1960: Notes on the genus *Aspro* Cuvier 1829 (Osteichthyes, Percidae). Acta Univ. Carolinae – Biologica **1:** 37–42.

OLIVER, G., 1970: Observations sur quelques poissons des eaux douces et saumâtres des Pyrénées-Orientales. Vie et Milieu, Ser. C. **(21):** 199–208.

PELZ, G. R., 1985: Fischbewegungen über verschiedenartige Fischpässe am Beispiel der Mosel. Cour. Forsch.-Inst. Senckenberg, 76: 190 pp., Frankfurt/M.

PEŇÁZ, M., P. RÁB & M. PROKEŠ, 1979: Cytological analysis, gynogenesis and early development of *Carassius auratus* gibelio. Acte Sc. Nat. Brno, **13** (7): 33 pp.

PIVNIČKA, K., and K. HENSEL, 1978: Morphological variation in the genus *Thymallus* Cuvier, 1829 and recognition of the species and subspecies. Acta Universitas Carolinae – Biologica 1975–1976: 37–67.

PRAWOCHENSKY, R., and W. KOLDER, 1968: Synopsis of biological data on *Hucho hucho* (Linnaeus, 1758). FAO Fisheries Synopsis No. **22** (Suppl. 1), Rome.

RAMEYE, L., A. KIENER, C.-P. SPILLMANN, and J. BIOUSSE, 1976: Aspects de la biologie de l'Alose du Rhône – pêche et difficultés croissantes de ses migrations. Bull. Franc. de Pisciculture, No. **263:** 50–76.

RANDIK, A. K. (Ed.), 1976: Proceedings of the Conference: The contemporary condition, protection and perspective of the Danube salmon in Slovakia (Czechoslovakia) and Central Europe. Priroda Press, Bratislava.

SCOTT, W. B., and E. J. CROSSMAN, 1973: Freshwater fishes of Canada. Fisheries Research Board of Canada, Bull. **184.**

SELLEY, L. Y., and F. W. H. BEAMISH, 1977: The Cyclostomata. An annotated bibliography. Dr. W. Junk, The Hague.

SIEBOLD, C., 1863: Die Süßwasserfische von Mitteleuropa. Leipzig.

SKROCHOWSKA, S., 1969: Migration of the sea-trout (*Salmo trutta* L.), brown trout (*Salmo trutta m. fario* L.) and their crosses. Part I to IV. Pol. Arch. Hydrobiol. **16** (29) 2: 125–192.

SMYLY, W. J. P., 1955: On the biology of the stone-loach *Nemacheilus barbatula* (L.). Journal Anim. Ecol. **24:** 167–186.

STANESCU, G., 1971: *Romanichthys valsanicola* Dumitrescu, Banarescu and Stoica (Pisces: Percidae), its distribution in Romania and the causes of its extinction. Acta soc. zool. Bohemoslov. **35:** 132–135.

STANKOVIC, S., 1960: The Balkan Lake Ohrid and its living world. Monographiae Biologicae, Vol. **19.** Dr. W. Junk, The Hague.

STEINMANN, P., 1950/1951: Monographie der schweizerischen Koregonen. Beitrag zum Problem der Entstehung neuer Arten. Schweiz. Z. Hydrologie **12:** 109–189, 340–491; 13: 54–155.

STRAŠKRABA, M., J. ČIHAŘ, S. FRANK, and V. HRUŠKA, 1966: Contribution to the problem of food competition among the sculpin, minnow and brown trout. J. Anim. Ecol. **35:** 303–312.

SVARDSON, G., 1957: The coregonid problem. VI. The Palearctic species and their intergrades. Rep. Inst. Freshw. Res. Drottningholm **38:** 268–356.

SVETOVIDOV, A. N., and E. A. DOROFEEVA, 1963: Systematics, origin and history of the distribution of the Eurasian and North American perches and pikeperches (genera *Perca, Lucioperca* and *Stizostedion*). Vopr. Ichtiol. **3:** 625–651 (Translated from Russian by NMFS Syst. Lab. No. 28).

TACK, E., 1940: Die Ellritze (*Phoxinus laevis* Ag.) eine monographische Bearbeitung. Arch. f. Hydrobiol. **37:** 321–425.

THORPE, I. E., 1977: Morphology, physiology, behaviour and ecology of *Perca fluviatilis* L. and *Perca flavescens* Mitchell. J. Fish. Res. Board. Can. **34:** 1504–1514.

TEROFAL, F., 1977: Das Artenspektrum der Fische Bayerns in den letzten 50 Jahren. Ber. ANL 1: 9–22; Laufen-Salzach.

TESCH, F.-W., 1972: Der Aal, Biologie und Fischerei. Paul Parey, Hamburg.

THIENEMANN, A., 1921: Über einige schwedische Coregonen, mit Bemerkungen über die Systematik der Gattung *Coregonus* und die Wege und Ziele der künftigen Coregonenforschung. Arch. f. Naturgeschichte **87:** 170–195.

THIENEMANN, A., 1922: Weitere Untersuchungen an Coregonen. Arch. Hydrobiol. **13:** 415–471.

THIENEMANN, A., 1941: Die Süßwasserfische Deutschlands, eine tiergeographische Skizze. Handbuch d. Binnenfischerei Mitteleuropas, Bd. **3a:** 1–32, Schweizerbart'sche Verlagsbuchhandlung, Stuttgart.

THIENEMANN, A., 1950: Verbreitungsgeschichten der Süßwassertierwelt Europas. Die Binnengewässer **18.** Schweizerbart'sche Verlagsbuchhandlung, Stuttgart.

THIENEMANN, A., 1955: Warum Gewässerschutz? Luxemburger Liga für Gewässerschutz, 16 pp.

THIENEMANN, A., 1956: Das Wasser als Sorge Europas, Universitas (Stuttgart) **11:** 849–858.

VARLEY, M. E., 1967: British freshwater fishes. Factors affecting their distribution. Fishing News, London.

VERHEY, C. J., 1949: Het vorkommen van den steur *(Acipenser sturio)* in de nieuwe Merwede tusschen 1900–1931. De Levende Natur **52:** 152–159.

VIBERT, R., 1957: Du rendement comparé des divers modes de repeuplement en truites et saumons. Bull. Cent. Etud. Rech. sci. Biarritz **3:** 321–342.

WAGLER, E., 1941: Die Lachsartigen (Salmonidae) II. Teil Coregonen. Handb. d.

Binnenfischerei Mitteleuropas **III A:** 371–501, Schweizerbart'sche Verlagsbuchhandlung, Stuttgart.

WEATHERLEY, A. H., 1963: Zoogeography of *Perca fluviatilis* (Linnaeus) and *Perca flavescens* (Mitchell) with special reference to the effects of high temperature. Proc. Zool. Soc. London **141:** 557–576.

WHEELER, A., 1969: The fishes of the British Isles and North-West Europe. Macmillan, London.

WHITE, R. J., 1968: So baut man Forellenunterstände. Paul Parey, Hamburg.

WHITE, R. J., and O. M. BRYNILDSON, 1967: Guidelines for management of trout stream habitat in Wisconsin. Dept. of Nat. Res., Div. of Conservation, Madison, Wisconsin. Techn. Bull., **39.** 65 pp.

ZSCHOKKE, F., and P. STEINMANN, 1911: Die Tierwelt der Umgebung von Basel. Helbing u. Lichtenhahn, Basel. (Ex Lauterborn 1918).

7. Index

7.1. Index of scientific names

Page numbers, set in bold faces indicate the beginning of the main section, those set in italics refer to distribution map.

7.2. Index of common names